T0190101

:r Buchreihe ist die Publikation neuer Ansätze in der Automation auf wissenschaft-
ı Niveau, Themen, die heute und in Zukunft entscheidend sind, für die deutsche und
ationale Industrie und Forschung. Initiativen wie Industrie 4.0, Industrial Internet
Cyber-physical Systems machen dies deutlich. Die Anwendbarkeit und der indust-
Nutzen als durchgehendes Leitmotiv der Veröffentlichungen stehen dabei im Vorder-
. Durch diese Verankerung in der Praxis wird sowohl die Verständlichkeit als auch
·levanz der Beiträge für die Industrie und für die angewandte Forschung gesichert.
Buchreihe möchte Lesern eine Orientierung für die neuen Technologien und deren
ndungen geben und so zur erfolgreichen Umsetzung der Initiativen beitragen.

ısgegeben von
- Institut für industrielle Informationstechnik
schule Ostwestfalen-Lippe
ɪo, Germany

Technologien für die intelligente Automation
Technologies for Intelligent Automatior
Band 1

Ziel
liche
inter
oder
rielle
grun
die F
Dies
Anw

Her
inIT
Hocl
Lem

Weitere Bände in dieser Reihe
http://www.springer.com/series/13886

Oliver Niggemann • Jürgen Beyerer (Eds.)

Machine Learning for Cyber Physical Systems

Selected papers from the international
Conference ML4CPS 2015

 Springer Vieweg

Editors
Oliver Niggemann
inIT - Institut für industrielle Informations-
technik
Hochschule Ostwestfalen-Lippe
Lemgo, Germany

Jürgen Beyerer
Fraunhofer-Institut für Optronik
Systemtechnik und Bildauswertung
Karlsruhe, Germany

Technologien für die intelligente Automation
ISBN 978-3-662-48836-2   ISBN 978-3-662-48838-6 (eBook)
DOI 10.1007/978-3-662-48838-6

Library of Congress Control Number: 2016931897

Printed on acid-free paper

Springer Vieweg is a brand of Springer Berlin Heidelberg
Springer Berlin Heidelberg GmbH is part of Springer Science+Business Media
(www.springer.com)

Preface

Data is the coming resource of the 21st century, e.g. the market capitalization of Google has already achieved almost the value of Exxon Mobil. In the future, this trend will continue for Cyber Physical Systems: E.g. globally connected production systems optimize automatically their energy consumption (keyword: Industrie 4.0), cars react dynamically to the driving behavior of other road users and trains detect wear effects beforehand.

This huge amount of generated data leads to completely new and unresolved challenges for data analysis and machine learning: McKinsey estimates that almost 2 Exabyte of new data were generated in the manufacturing industry in 2010. The amount of data prohibits any manual analysis, e.g. by classical data scientists.

The solution can only be the use of highly automated machine learning methods. But most of these methods do not consider peculiarities of technical systems: The dynamic time behavior is not modeled, control signals and the resulting behavior changes of hybrid systems are not captures and physical knowledge is not used.

Therefore the conference ML4CPS aims at bringing experts from science and industry together to discuss current demands on machine learning for Cyber Physical Systems and match them with recent results from the scientific community.

Prof. Dr. rer. nat. Oliver Niggemann   Prof. Dr.-Ing. habil. Jürgen Beyerer

Contents

Development of a Cyber-Physical System based on selective Gaussian naïve Bayes model for a self-predict laser surface heat treatment process control

Javier Diaz[1], Concha Bielza[2], Jose L. Ocaña[3], and Pedro Larrañaga[2]

[1]Ikergune A.I.E, San Antolin 3, 20870 Elgoibar, Spain
jdiaz@ikergune.com
[2]Department of Artificial Intelligence, Technical University of Madrid, School of
Computer Science. Madrid, Spain
[3]UPM Laser Centre, Technical University of Madrid. Madrid, Spain

Abstract. Cyber-Physical Systems (CPS) seen from the Industrie 4.0 paradigm are key enablers to give smart capabilities to production machines. However, close loop control strategies based on raw process data need large amounts of computing power, which is expensive and difficult to manage in small electronic devices. Complex production processes, like laser surface heat treatment, are data intensive, therefore, the CPS development for these type of processes is challenging. As a result, the work described in this paper uses machine learning techniques like naïve Bayes classifiers and feature selection optimization, in order to evaluate its performance during surface roughness detection. Additionally, the feature selection techniques will define optimal measuring zones to reduce generated data. The models are the first step towards its future embedding into a laser process machine CPS and bring *self-predict* capabilities to it.

Keywords: CPS, machine learning, naïve Bayes, laser heat treatment, self-predict, Industrie 4.0

1 Introduction

Nowadays, laser applications are growing in many sectors like medicine, metrology, telecommunications and industrial applications [1]. However, industrial applications represent near 63% of the global market ($1.671 million last year), where high-power lasers used for macro-processing are the most interesting due to their potential of applicability in the manufacturing industry. Near 75% of these applications are related to sheet metal cutting and the remaining 25% is related with other processes like drilling or heat treatment [2].

As result of this, laser technologies have been evolving to allow their applicability in the manufacturing sector, where robustness, repeatability and reproducibility are key factors. Accordingly, process monitoring technologies have been developed in order to enable in-process quality control.

For a laser surface heat treatment process control, the main objective is to regulate the amount of energy deposited over the treated surface by the laser beam source [1]. The aim of this energy addition is to increase surface temperature and modify the mechanical properties of the surface, the physical phenomena has being essentially thermal [3] [4] [5].

Basically, laser surface quenching occurs when a high-density laser beam is focused on a small area where it causes the heating. In the case of steels, this involves its temperature to the austenite region. When the laser beam is moved, the area is immediately quenched by heat conduction to the surrounding metal [6].

Because of the thermal behaviour, laser process monitoring systems are mainly based on temperature reading, where contactless pyrometers and thermal cameras are often used. Additionally, if the monitoring systems seek to ensure the in-process quality, with low latency control loop, pyrometers are the most used option [7].

However, in a surface heat treatment the process area is larger than the area covered by the pyrometer, that increases the risk of losing critical information to effectively control the process. Costly analysis processes have to be carried out in order to estimate the optimal measuring point. Particularly, in the process of this study, the treated area is more than fifteen times the pyrometer range, so that trying to cover the complete area adding pyrometers is unmanageable.

As a result of the above mentioned problem, new developments towards the use of thermal cameras and image processing tools to cover wider areas are found in the literature [8]. Basically, these developments are based on thermal image processing in order to obtain a matrix of temperatures over the treated surface, which can be analysed and used as a matrix of pseudo-pyrometers.

Nonetheless, the control of the process based on this technology has an important drawback regarding the latency of the system. Due to the large amount of information analysed every process cycle, computing power may compromise the latency of the system, which is directly related to the size of the image and sampling time. This high latency prevents the use of this control strategy for in-process quality control.

Therefore, the objective of this work is to take advantage of cyber-physical systems (CPS) capability to handle large amounts of information integrated with processing capabilities [9] [10] [11] that will reduce the need of computing power to decrease the latency of a thermal camera control process. Additionally, using available monitoring technologies (in this case thermal cameras) complies with some Industrie 4.0 basic recommendations which explains that is not about putting more sensors inside processes [12] [13], but about extracting information from the actually available systems.

As a result of this latency reduction, the monitoring system will have in-process quality control capabilities. Machine learning techniques are used to identify patterns inside the thermal image during the process that can be classified in order to detect fluctuations during the process. This fingerprint of each fluctuation will be used to develop predictive models to control the laser power without the need of a complete pseudo-pyrometer matrix analysis every sampling time.

2   Methodology

Because laser beam energy deposition efficiency is strongly related to surface roughness due to its reflectivity, process parameters have to be adjusted. That is, lower roughness produces higher reflectivity which means that less laser energy is absorbed. Therefore, one of the main purposes of heat treatment process control is to be able to give the required amount of energy regardless of the surface roughness.

Because of this, for experimentation two scenarios arise: ground and unground workpiece surfaces. Consequently, machine learning techniques are used to classify the workpiece surface type within an acceptable limit of time. This time will have a strong influence over the laser beam control in order to modify the power input to correct the energy needs to achieve the optimal process temperature.

Therefore, the experimental setup is oriented to gather real data from a laser heat treatment process and evaluate the machine learning techniques performance in order to enable their application within a CPS device. Accordingly, data acquisition system and machine learning model are explained in the following sections.

2.1   Laser Heat Treatment Data Acquisition

As mentioned before, the laser heat treatment parameter to control the process is the amount of energy deposited over the surface, i.e., the process is mainly thermal. In a real control strategy of a laser process, the power needed from the laser beam source to achieve the heat treatment temperature is the parameter to be controlled, and, the quantitative input variable needed is the treated surface temperature during the process.

In order to gather the surface temperature, experimentation has been carried out over 12 real workpieces (6 ground and 6 unground) that are simplified as one set of cylinders with dimensions described in Table 1. Each set contains 4 cylinders type M and 6 cylinders type P. In order to get the process temperature, a FLIR A655sc 25° thermal camera with 640 x 480 pixels at 25/12.5 Hz has been used to record each cycle. The cycle time is between 20 to 25 seconds, depending on the cylinder size.

In order to extract the temperature data from each frame, an image processing software, developed in collaboration with Vicomtech-IK4, has been used to deploy over the laser heated zone a 3 x 5 matrix of *virtual* pyrometers. The treatment zone size is 20 x 10 mm. Therefore, the measurements taken from the *virtual* pyrometer array are stored in a CSV file for further manipulation. The complete process is described in Figure 1.

Table 1. Cylinder sizes and quantities

Cylinder Type	Diameter [mm]	Height [mm]	Ground Surface Cyl.	Unground Surface Cyl.	Total Cyl.
M	67.85	19.19	24	24	48
P	56.60	15.24	36	36	72

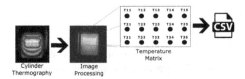

Fig. 1. Data acquisition process.

2.2 Machine Learning Experimental Setup

From a machine learning viewpoint, for each type of cylinder described in Table 1, a naïve Bayes model will classify its surface type based on the information given by the variables, i.e. temperature values, contained in each thermography frame. Therefore, the percentage of correctly and incorrectly classified instances will be used as a quantitative measurement for the classifier performance.

A *naïve Bayes classifier* is the simplest Bayesian classifier. It is built upon the assumption of conditional independence of the predictive variables given the class. Although, this assumption is violated in numerous occasions in real domains, the paradigm still perform well in many situations. The most probable *a posteriori* assignment of the class variable is calculated as prediction.

$$c^* = arg\max_c p\left(c|x_1,\ldots,x_n\right) = arg\max_c p(c)\prod_{i=1}^{n} p(x_i|c)$$

Consequently, in order to evaluate the minimum time required by the classifier to detect the surface type of the cylinder, an incremental amount of frames will be given to the model. Each frame represents a specific amount of time, depending on the video speed used by the thermal camera, i.e. 12.5 Hz is equivalent to 80 ms and 25 Hz is equivalent to 40 ms. Therefore, if the classifier needs 1 frame to reach the desired accuracy, it will mean that the model can give a result in 80 ms at 12.5 Hz. As a result, experiments have been carried out for cylinders Type M and P from 1 frame to 20 frames, which is between 0.8 and 1.6 s depending on the frame rate used in the camera.

After running experiments for each type of cylinder, 3 different types of feature selection (FS) techniques have been carried out to increase the classifier accuracy. These FS techniques will assist classifier models to deal with large amounts of irrelevant information [14].

Therefore, *filter techniques* assess the relevance of the variables, in this case, the temperatures inside the defined matrix, by looking only at the intrinsic properties of the data. In this experimentation, two types of filter have been used: Correlation-based feature selection (CFS) was introduced by [15]. CFS seeks for a good feature subset, that is one that contains features highly correlated with the class, yet uncorrelated each other. More formally, denoting by $\mathcal{S}$ a subset of the predictive features $\mathcal{X}$, CFS looks for $\mathcal{S}^* = \arg\max_{\mathcal{S}\subseteq\mathcal{X}} r(\mathcal{S}, C)$, where

$$f(\mathcal{S}) = r(\mathcal{S}, C) = \frac{\sum_{X_i \in \mathcal{S}} r(X_i, C)}{\sqrt{k + (k-1)\sum_{X_i, X_j \in \mathcal{S}} r(X_i, X_j)}}$$

measures the correlation between the selected features and the class variable, k is the number of selected features, $r(X_i, C)$ is the correlation between feature X_i

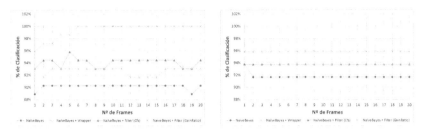

Fig. 2. Classifier accuracy for Type P (left) and M (right).

and the class variables C, and $r(X_i, X_j)$ is the inter-correlation between features X_i and X_j.

The other type of filter used is Information Gain Attribute Evaluation (Gain Ratio). The *information gain* between two variables X_j and C:

$$f(X_j) = \mathbb{I}(X_j, C) = -\sum_{i=1}^{r_j} \sum_{c=1}^{r_0} p(x_i, c) \log_2 p(x_i, c)$$

measures the reduction in the uncertainty of one of them (variable C for example) once the value of the other variable (X_j for example) is known. For Type P test-pieces, the ranked level threshold selected is 0.40, which means that all variables below this value are removed. For Type M, the selected ranked level is 0.92. The thresholds where selected based on the maximum value where the accuracy of the classifier starts to be negatively affected.

Additionally, *wrapper methods* [16] evaluate each possible subset of features with a criterion based on the estimated performance of the classifier built with this subset of features.

The complete experimentation for classification has been run over Weka, which is a collection of machine learning algorithms for data mining [17].

3   Results

The classifier accuracy for Type P cylinders is shown in Figure 2 (left), with and without feature selection techniques applied. For experiments with the naïve Bayes classifier, with only 1 frame the model is able to achieve near 89% of accuracy. Then, with 2 frames and more, the classifier is stable at 90.2% accuracy. This means that the classifier will be able to detect ground and unground surfaces in 160 ms with an accuracy of 90.2%.

On the other hand, for Type M cylinders shown in Figure 2 (right), the naïve Bayes classifier alone achieves 93.8% for 1 frame, decreasing to 91.6% for 2 frames or more. This situation shows a better behaviour than Type P cylinders, getting better accuracy with the same time needed.

However, applying FS techniques, the classifier performance increases for both types of test-pieces, except for CFS in Type M. Regarding the results for Type P for naïve Bayes plus Gain Ratio filter, the accuracy using 1 frame is 93% which remains stable with more frames, expect for 2 frames and from 12 to 15 frames. Nevertheless, using Gain Ratio for Type M, increases the accuracy to

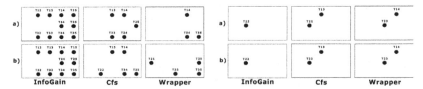

Fig. 3. Selected temperatures for a) one frame and b) two frames. Type P (left) and M (right).

95.8% remaining stable until 3 frames. These results are similar to naïve Bayes filtered with CFS technique.

Additionally, using CFS in Type P, the accuracy is similar to Gain Ratio for 1 frame. However, there is an improvement to 94.4% when 2 frames are used. In this case, the selected variables are reduced to 5 instead of 10 with Gain Ratio for 1 frame and 5 instead of 21 for 2 frames.

For Type M workpieces, the same accuracy using 1 frame is found for CFS as the classifier without selected features, nonetheless, it remains stable at 93.75% of accuracy independently of the number of frames given. Consequently, the number of selected variables is 2 compared to 1 using Gain Ratio for both: 1 and 2 frames.

On the other hand, using wrapper technique in Type P, the accuracy is relatively higher than filter techniques, having near 96% of accuracy for 1 frame increasing to 97% for 2 frames. It is important to stress that from 6 frames, the classifier is able to reach 100% of accuracy, meaning that from 6 frames up, the problem is linearly separable. In Type M, the accuracy is the same as Gain Ratio technique, outperforming other FS techniques from 3 frames.

The selected variables positioned within the pyrometers matrix, for 1 and 2 frames, are shown in figure 3.

4   Discussion

As seen in the previous section, feature selection techniques are able to increase the classifier accuracy in average by 9% for Type P and 4% for Type M. In general, the naïve Bayes classifier with wrapper has the best behaviour regarding its accuracy for both scenarios. Nevertheless, the comparison between models and the physical position within the matrix of the selected variables can bring more information regarding the physical behaviour of the temperature. This is relevant in order to develop an accurate but fast control system for the laser surface treatment.

As a result, in Figure 3 where the *virtual* pyrometers within the frame matrix are shown temperature measurement positions are selected at least by two techniques independently of the number of frames used, for example, T14, T34 or T35 for Type P and T22 in Type M. It is understandable, that those positions are relevant for the process as they are always selected.

From these positions, from the process point of view, that right leading (T14) and right trailing edges (T34 and T35) of the laser spot, in Type P scenario, are more sensitive to surface roughness variation which is expressed by an emitted temperature difference easier to detect by the classifier.

Nevertheless, for Type M, the most sensitive point is near the center of the matrix. This situation is interesting due to the fact that the only difference between each test-piece is its size, as process parameters are exactly the same for Type P. However, the number of selected temperature positions are significantly lower.

Therefore, it is expected that selected variables are the best descriptors of temperature differences between ground and unground surfaces. Therefore, in order to increase CPS ability to process information as fast as possible, reducing the amount of information given to it, the temperature measurements should be carried out in zones where best describers are located. Moreover, only pixels of interest within the frame can be gather, transmitted and processed, reducing computing power needs.

In the other hand, the time that the classifiers needs to detect the difference between temperature patterns at an acceptable accuracy is adequate, being always below 160 ms. However, this value depends on the thermal camera frame rate, so that, it is expected that data needed to classify might be obtained in less time, increasing the camera frame rate. In a control model, this situation would reduce the CPS latency.

5   Conclusion

The main conclusion of this paper is that the application of machine learning into CPS is key enabler to give decision capabilities to the analysed kind of devices. Those decisions can be implemented to the control system, adding self-adaptation capabilities to a real production process. In this case, high accuracy levels obtained from classifiers, even without feature selection, are able to detect differences between surface roughness, key condition for *optical* processes like laser heat treatments.

This experimental work only has provided results for two types of surface roughness. However, it is one of the most difficult boundary condition to be detected by machine operators due to small differences between temperatures. In this case, the model only needs information at least from the first two frames of process thermography, which, as explained before should be reduced if the thermal camera speed is higher. This situation opens new application boundaries to integrate machine learning models to control laser surface treatment process.

Therefore, because of the positive results found in this work, further experimentation on the application of machine learning for a cyber-physical system that controls laser surface heat treatment is oriented to detect and adapt the process to different surface roughness values. In this case, due to the results obtained, the classifier model will be able to relate surface reflectivity (directly related to roughness) and temperature values gathered by thermography.

Other applications for future investigation enabled by this work are to detect and alert about surface crack generation and monitor laser scanning system mechanical degradation to give the process predictive maintenance capabilities, all of them part of what is considered, a Cyber-Physical System.

Acknowledgments. This work has been supported by Spanish Centre for the Development of Industrial Technology (CDTI) through TIC-20150093 and it has been partially supported by the Spanish Ministry of Economy and Competitiveness through TIN2013-41592-P project and by the Regional Government of Madrid through the S2013/ICE-2845-CASI-CAM-CM project. Authors thank Vicomtech-IK4 providing developments for image processing software used for this work.

References

1. Pérez, J.A., López, M.: Design and implementation of an innovative quadratic gaussian control system for laser surface treatments. The International Journal of Advanced Manufacturing Technology **65**(9-12) (2013) 1785–1790
2. Belforte, D.: Fiber lasers continue growth streak in 2014 laser market. `http://goo.gl/z2nXkH` (2015) [Online; accessed 07-July-2015].
3. Pérez, J.A., Ocana, J.L., Molpeceres, C.: Hybrid fuzzy logic control of laser surface heat treatments. Applied Surface Science **254**(4) (2007) 879–883
4. Badkar, D.S., Pandey, K.S., Buvanashekaran, G.: Parameter optimization of laser transformation hardening by using taguchi method and utility concept. The International Journal of Advanced Manufacturing Technology **52**(9-12) (2011) 1067–1077
5. Pérez, J.A., Ocana, J.L., Molpeceres, C.: Real time fuzzy logic control of laser surface heat treatments. In: Industrial Electronics. ISIE. IEEE International Symposium on, IEEE (2007) 42–45
6. Groover, M.P.: Fundamentals of Modern Manufacturing: Materials Processes, and Systems. John Wiley & Sons (2007)
7. Qian, B., Taimisto, L., Lehti, A., Piili, H., Nyrhilä, O., Salminen, A., Shen, Z.: Monitoring of temperature profiles and surface morphologies during laser sintering of alumina ceramics. Journal of Asian Ceramic Societies **2**(2) (2014) 123–131
8. Thombansen, U., Gatej, A., Pereira, M.: Process observation in fiber laser–based selective laser melting. Optical Engineering **54**(1) (2015) 011008–011008
9. Kagermann, H., Helbig, J., Hellinger, A., Wahlster, W.: Recommendations for Implementing the Strategic Initiative INDUSTRIE 4.0: Securing the Future of German Manufacturing Industry; Final Report of the Industrie 4.0 Working Group. Forschungsunion (2013)
10. Baheti, R., Gill, H.: Cyber-physical systems. The Impact of Control Technology **12** (2011) 161–166
11. Babiceanu, R.F., Seker, R.: Manufacturing cyber-physical systems enabled by complex event processing and big data environments: A framework for development. In: Service Orientation in Holonic and Multi-agent Manufacturing. Springer (2015) 165–173
12. Diaz, J., Posada, J., Barandiaran, I., Toro, C.: Recommendations for sustainability in production from a machine-tool manufacturer. KES-SDM (2015) In press.
13. Posada, J., Toro, C., Barandiaran, I., Oyarzun, D., Stricker, D., de Amicis, R., Pinto, E.B., Eisert, P., Döllner, J., Vallarino, I.: Visual computing as a key enabling technology for industrie 4.0 and industrial internet. IEEE Computer Graphics and Applications **35**(2) (2015) 26–40
14. Saeys, Y., Inza, I., Larrañaga, P.: A review of feature selection techniques in bioinformatics. Bioinformatics **23**(19) (2007) 2507–2517
15. Hall, M.: Correlation-based Feature Selection for Machine Learning. PhD thesis, Department of Computer Science, University of Waikato (1999)
16. Kohavi, R., John, G.: Wrappers for feature subset selection. Artificial Intelligence **97**(1) (1997) 273–324
17. Hall, M., Frank, E., Holmes, G., Pfahringer, B., Reutemann, P., Witten, I.H.: The weka data mining software: An update. SIGKDD Explorations **11**(1) (2009)

Evidence Grid Based Information Fusion for Semantic Classifiers in Dynamic Sensor Networks

Timo Korthals[1], Thilo Krause[2], and Ulrich Rückert[1]

[1] Bielefeld University, Cognitronics & Sensor Systems,
Inspiration 1, D-33619 Bielefeld, Germany
{tkorthals, urueckert}@cit-ec.uni-bielefeld.de
[2] CLAAS Selbstfahrende Erntemachinen GmbH, Münsterstr. 33,
D-33428 Harsewinkel, Germany
thilo.krause@claas.com

Abstract. We propose an anytime fusion setup of anonymous distributed information sources with spatial affiliation. For this approach we use the evidence grid mapping algorithm which allows to fuse sensor information by their inverse sensor model. Furthermore, we apply an online Mixture of Experts training such that faulty voters are detected and suppressed during runtime by a gating function.

Keywords: Evidence grid maps, sensor network, faulty sensor detection

1 Introduction

We are concentrating on a cyber-physical system as a realization of a mobile working vehicle acting on its environment. The main requirements originates from a harvesting application which comprises the exploration and mapping within a dynamic setup of sensors and vehicles. One requirement is the acquisition of exteroceptive conditions like traversability, machinability, and applicability of the crop to be processed. Another belongs to the sensory setup with which the information acquisition is done. It substantially differs from standard vehicle sensor setups like automobiles, due to their changeable operating conditions which is caused by different goods, unstructured or rough environment, or simply the demand of different sensor setups.

In particular our setup is a production system equipped with a distributed set of dynamic heterogeneous intelligent sensors (IS) which comprises but is not limited to cameras, laser range finders, and proximity sensors at position x which are able to map a measurement z to a distinctive location in space m. Every IS can be an own embedded system which is a set of a physical sensor and a processing system consisting of the driver for interfacing the raw data and a set of classifiers which converts the raw data into an evidence grid representing a common source of information (see left fig. 1). We call this information "semantic feature" due to the fact that the semantic meaning, e.g. the representation of

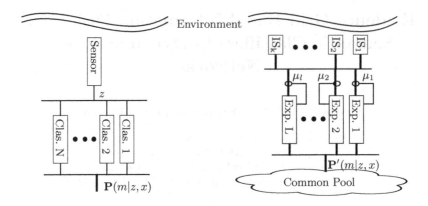

Fig. 1: Left: Intelligent sensor with multiple classifiers. Right: Local expert system.

a stock boundary, is always attached to it. Furthermore, we are not only concentrating on an information fusion system given a single static sensor setup, but a versatile one where sensors can be added or removed even during runtime.

To meet these requirements, our setup is build up as can be seen in the right of fig. 1: Different IS are distributed all over the machinery which may overlap with their detection ranges. All IS can anonymously distribute their evidence grids $\mathbf{P}\left(m|z,x\right) = \left(\mathrm{P}_1\left(m|z,x\right),\ldots,\mathrm{P}_N\left(m|z,x\right)\right)^{\mathrm{T}}$ into the network where local expert systems listen to their corresponding semantic features. The local expert then fuses a single grid by Elfes [3] occupancy grid map algorithm in a normalized fashion and backpropagates the information such, that faulty classifiers are going to be suppressed using a single parameter μ. The fused information $\mathbf{P}'\left(m|z,x\right) = \left(\mathrm{P}_1\left(m|z,x\right),\ldots,\mathrm{P}_L\left(m|z,x\right)\right)^{\mathrm{T}}$ is send to the common pool, from which further experts or remote machines get the experts outcome for further reasoning.

The paper is organized as follows: First, we introduce the evidence grid framework in chapter 2. Second, the enhancement of the fusion formulation by the gating network is explained in chapter 3. Finally, we evaluated our approach using a robotic simulation in a standard exploration task in 4 and give a brief summary and discussion in 5.

2   Evidence Grid Map Algorithm

In this section we will discuss how a spatial representation of the presence of a property can be learned via an evidence grid map from any sensor data using Bayesian update rules. First introduced by Moravec [8], evidence grids maps are currently the state of the art approach for enabling navigation of mobile robots in real world environments [4–6]. The environment is discretized using an evenly spaced grid, with each cell holding a probabilistic estimate of its property. For the sake of simplicity, we do the derivation for a single expert l $(m \equiv m_l)$ as we assume independency between all representations. Given the positions x_{t_n} of IS n at each point in time t is known, where n is drawn from the set of sensors. Suppose $x_{1_n:t_n} = x_{1_n}, \ldots, x_{t_n}$ are the positions of any IS n at the individual steps in time and $z_{1_n:t_n} = z_{1_n}, \ldots, z_{t_n}$ are the perceptions of the environment regarding an IS. Evidence probability grids determine for each cell m of the grid the probability that this cell is occupied by a given representation. Thus, evidence probability grids seek to find the map M that maximizes $\mathrm{P}(m|x_{1_n:t_n}, z_{1_n:t_n})$ for each cell. If we apply Bayes rule using $x_{1_n:t_n}$ and $z_{1_n:t_n}$ as background knowledge, we obtain

$$\mathrm{P}(m|x_{1:t}, z_{1_n:t_n}) = \frac{\mathrm{P}(z_t|m, x_{1_n:t_n}, z_{1_n:t_n-1})\mathrm{P}(m|x_{1_n:t_n}, z_{1_n:t_n-1})}{\mathrm{P}(z_t|x_{1_n:t_n}, z_{1_n:t_n-1})}. \quad (1)$$

We assume that z_{t_n} is independent from $x_{1_n:t_n-1}$ and $z_{1_n:t_n-1}$ given we know m, the right side of this equation can be simplified such that n and $z_{1:t-1}$ can be omitted. This results to the equation for a single sensor as introduced by Moravec [8]:

$$\mathrm{P}(m|x_{1:t}, z_{1:t}) = \frac{\mathrm{P}(z_t|m, x_t)\mathrm{P}(m|x_{1:t}, z_{1:t-1})}{\mathrm{P}(z_t|x_{1:t}, z_{1:t-1})}. \quad (2)$$

Eq. 2 leads us to an update rule to incorporate a new scan into a given map by multiplying its odds ratio $\mathrm{R} = \mathrm{P}/(1-\mathrm{P})$ with the odds ratio of a local map constructed from the most recent scan, divided by the odds ratio of the prior $\mathrm{R}(m)$. Often it is assumed that the prior probability of m is 0.5 (s.t. unknown). In this case the prior can be cancelled out, writing the log odds representation as follows:

$$\log \mathrm{R}(m|x_{1:t}, z_{1:t}) = \log \mathrm{R}(m|x_t, z_t) + \log \mathrm{R}(m|x_{1:t-1}, z_{1:t-1}). \quad (3)$$

Eq. 3 tells us how to update our belief about the occupancy probability of a grid map given the current sensory input z_t with $\log \mathrm{R}(m|x_t, z_t)$ being the log odds representation of the so called inverse sensor model $\mathrm{P}(m|x_t, z_t)$ which can be derived empirically or trained by any gradient descent technique [10].

As a rule of thumb the input probabilities as well as the output probabilities are bounded to $\mathrm{P} \in (0.2, 0.8)$ to avoid over fitting and to retain a good dynamic range when new data arise. A beneficial effect of the bounding is the fact that the log odds ratio becomes a function f which maps P almost linearly into the real numbers. With respect to the task we take a distinctive set of measurements comprising the last time interval T. This technique has first be proposed by Arbuckle [1] to assure that also dynamic and quasi static objects can be mapped by the grid mapping algorithm. Therefore, we concentrate on this particular measurement set of timespan T:

$$\log \mathrm{R}(m|x_T, z_T) \approx \sum_{t \in T} f(\mathrm{P}(m|z_t, x_t)). \quad (4)$$

3   Mixture of Experts

Mixture of Experts (ME) models attempt to solve problems using a divide-and-conquer strategy. They learn to decompose complex problems into simpler subproblems, that is the expert network to partition the input space onto different voters and an intelligent sensors (IS) network which is responsible for the particular region and semantic feature.

Our approach is the application of a gating function which has the property of scaling each inverse sensor model at the particular cell m such that faulty voters are going to be suppressed. Eq. 4 is in a form to fit well in a Mixture of Experts model

$$\log R(m|x_T, z_T) \approx \sum_{t \in T} g(z_t, x_t) f(P(m|z_t, x_t)) \qquad (5)$$

with $g(z, x)$ being the gating function μ. In eq. 5 the gating function itself is the only one which can be optimized due to the fact, that the inverse sensor models are given. Thrun et. al. [10] proposed the learning of the inverse sensor model by a neuronal network. In that case, the whole system can be trained in a supervised fashion as commonly known for ME models [7].

Within this work the gating function is the soft max activation which is trained for each timespan T by the error function

$$E = \sum_{t \in T} \|y(m) - g(z_t, x_t) r(m|z_t, x_t)\| . \qquad (6)$$

$y(m), r(m|z_t, x_t) \in \{-1, 1\}$ are the overall and currently classified binary classifications of the cell m of being occupied by a feature.

4   Evaluation

We implemented our proposed framework into the AMiRo robot developed by Herbrechtsmeier et. al. [9]. Three different sensors with spatial coverage shown in fig. 2 were used, where each one is attached with a different application creating the inverse sensor model for the semantic feature "obstacle": A LIDAR was used in combination with Thruns [10] proposed obstacle detection (IS1). Second, a set of eight cocircular arranged proximity sensors were used to detect close obstacles in combination with a detection by Benet et. al. [2] (IS2). At last, we applied a WXGA camera to detect anomalies using the proposed algorithm by Kohlbrecher et. al. [5]. We used the GazeboSim simulator shown in fig. 3 for the robot and created an occupancy map of obstacles in the environment for an exploration task. The coverage of the map compared to the ground truth has been measured over a simulated false-negative rate for a given sensor. Assuming an ideal IS, the false-negative rate is the probability of a feature not being detected by an IS. As long as the false-negative rate of a sensor is below 50%, the evidence grid framework converges to the correct solution anyway due to the fact, that the fusion corrects the measurements over multiple readings. On the other hand, if the rate is going to be above 50% the regions which cannot be correctly classified by the non-faulty voters will be detected wrong, resulting in a massive coverage failure. As can be seen in tab. 1, the gating network detects faulty classifiers

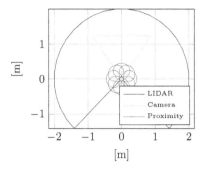

Fig. 2: Full range of sensor cones   Fig. 3: Simulator view

Fig. 4: Learned spatial gating functions for IS1 (left), IS2 (middle), and IS3 (right). Grayscale represents the gating function value from 0 (white) to 1 (black).

Table 1: Mean coverage until convergence of the mapping process of the obstacle feature (Standard deviation over multiple runs < .1%). Coverage without/with gating and no faulty sensor is both .81.

	Affected Sensor	Coverage				
False-Negative Rate		20%	40%	60%	80%	100%
Without Gating	IS1	.80	.79	.22	.21	.19
	IS2	.80	.81	.30	.22	-
	IS3	.78	.76	.22	.19	.14
With Gating	IS1	.80	.80	.76	.75	.75
	IS2	.80	.80	.79	.79	.79
	IS3	.78	.77	.70	.70	.69

as soon as their majority of measurements is faulty. The gating function for each IS is learned as a spatial representation as shown in fig. 4 for a non-faulty case. The exploration highly rely on the proximity sensors for the near field environment detection. While having a false-negative rate of 100%, meaning that objects are missed constantly, the robot is not able to detect objects for collision avoidance and thus the exploration fails. Equipped with the gating framework, we let the robot move by chance in the beginning which results in the detection and exclusion of the faulty proximity sensors.

5   Summary

To meet the trend of cyber-physical systems applications, where physical processing systems are equipped with sensors attached to communication networks, we enhanced the evidence grid framework to fuse spatial and semantic information of distributed, heterogeneous sensors. First, we designed every sensor system as an IS such, that it emits its information as an inverse sensor model with semantic affiliation attached. Then, we fused the information of a set of IS by applying the computational efficiency log odds update rule. Using our proposed enhanced fusion, the results correctly converge in our simulated robot exploration task even if faulty IS exists due to the application of a gating term to every fusion expert. While every single instance, s.t. IS, expert, or further information processing applications, is independent and not bounded to one physical system, the whole framework suits the demands of cyber-physical systems. In future work we will apply our framework to heterogeneous robot and harvesting scenarios with further cognitive processes acting on the fused information.

Acknowledgement. This research and development project is funded by the German Federal Ministry of Education and Research (BMBF) within the Leading-Edge Cluster "Intelligent Technical Systems OstWestfalenLippe" (it's OWL) and managed by the Project Management Agency Karlsruhe (PTKA). The author is responsible for the contents of this publication. This research was supported by the Cluster of Excellence Cognitive Interaction Technology 'CITEC' (EXC 277) at Bielefeld University, which is funded by the German Research Foundation.

References

1. Arbuckle, D., Howard, a., Mataric, M.: Temporal occupancy grids: a method for classifying the spatio-temporal properties of the environment. IEEE/RSJ International Conference on Intelligent Robots and Systems 1, 409–414 (2002)
2. Benet, G., Blanes, F., Simó, J., Pérez, P.: Using infrared sensors for distance measurement in mobile robots. Robotics and Autonomous Systems 40(4) (2002)
3. Elfes, A.: Dynamic control of robot perception using multi-property inference grids (1992)
4. Haehnel, D.: Mapping with Mobile Robots. Ph.D. thesis (2004)
5. Kohlbrecher, S.: Grid-based occupancy mapping and automatic gaze control for soccer playing humanoid robots. . . . Humanoid Soccer Robots . . . (October) (2011)
6. Li, Y., Ruichek, Y.: Occupancy grid mapping in urban environments from a moving on-board stereo-vision system. Sensors (Switzerland) 14(6), 10454–10478 (2014)
7. Masoudnia, S., Ebrahimpour, R.: Mixture of experts: A literature survey. Artificial Intelligence Review 42(2), 275–293 (2014)
8. Moravec, H., Elfes, a.: High resolution maps from wide angle sonar. Proceedings. 1985 IEEE International Conference on Robotics and Automation 2 (1985)
9. Stefan Herbrechtsmeier, Ulrich Rückert, J.S.: AMiRo - Autonomous Mini Robot for research and education. Springer Berlin Heidelberg, Berlin, Heidelberg (2012)
10. Thrun, S., Burgard, W., Fox, D.: Probabilistic Robotics. Intelligent robotics and autonomous agents, MIT Press (2005)

Forecasting Cellular Connectivity for Cyber-Physical Systems: A Machine Learning Approach

Christoph Ide, Michael Nick, Dennis Kaulbars, and Christian Wietfeld

Communication Networks Institute, TU Dortmund University, Germany
{christoph.ide,michael.nick,dennis.kaulbars,
christian.wietfeld}@tu-dortmund.de
http://www.kn.e-technik.tu-dortmund.de/en/

Abstract. Many applications in the context of Cyber-Physical Systems (CPS) can be served by cellular communication systems. The additional data traffic has to be transmitted very efficiently to minimize the interdependence with Human-to-Human (H2H) communication. In this paper, a forecasting approach for cellular connectivity based machine learning methods to enable a resource-efficient communication for CPS is presented. The results based on massive measurement data show that the cellular connectivity can be predicted with a probability of up to 78 %. Regarding a mobile communication system, a predictive channel-aware transmission based on machine learning methods enables a gain of 33 % concerning the spectral resource utilization of an Long Term Evolution (LTE) system.

1 Introduction

Individual human mobility has a clearly noticeable impact on the connectivity of a mobile communication system. Connectivity represents the quality of a transmission channel and is influenced by location-dependent effects, e.g., pathloss or shadowing effects in urban canyons. This paper copes with a connectivity forecast in mobile communication systems based on machine learning methods. Based on recent past-values of a connectivity indicator, a change of future connectivity can be predicted by applying them on a classification model by neglecting any tracking of position data. A connectivity indicator represents a time-dependent channel-quality, whereas its temporal signal course contains knowledge about the individual characteristic of a certain spatial route. A classification model is trained on massive data, either measured experimentally in a real scenario or computed in a simulative reference scenario. Therefore, descriptive attributes on recent past-values are mapped with a label describing a future change of connectivity. The connectivity forecast based on machine learning methods is used for a predictive channel-aware transmission that leverages the predicted connectivity hot spots for a CPS data transmission.

2   Related Work

The vehicular traffic is predictable in many cases. This is due to the regularity of the user behavior. It is analysed in [1] that 93% of the vehicular traffic is predictable. These results are based on a large study of 45,000 users. Other studies analyzed that 82% of the user mobility in Beijing and 89% in Shanghai [2] is predictable. This regularity can be used to forecast the mobile connectivity or to provide other services for mobile networks. In [3], machine learning methods are used to optimize handover processes in cellular communication systems based on mobility forecasts. The mobile performance can also be forecasted based on a connectivity map. This is particularly interesting for vehicular applications, because the mobility of cars make it possible to collect connectivity information at very different positions. In [4], a system that collects information about the mobile connectivity from different sources is presented. These sources include passive measurement results as well as position data from vehicles and from the cellular network. All these information are aggregated in a geo-database that includes the connectivity map. In addition, a connectivity forecast of cellular and WLAN systems by means of a Markovian model is shown in [5]. Thereby, each state represents a location with a previously measured throughput and the transitions between the states are calculated based on historical mobility information.

3   Predictive Channel-Aware Transmission (pCAT) based on Data Mining

A data transmission in a LTE mobile communication system at comparatively good channel quality can be performed very efficiently due to the possibility to use high modulation and coding schemes [6]. For this purpose, in CPS it is possible to delay sending processes of non-time-critical data to moments of favorable channel conditions of the underlying mobile communication channel as then the limited radio resources can be used more efficiently (cf. [7][8]). Formally expressed, particularly non-time-critical background-data are considered a sending decision that can depend on a transmission probability $p_T(t)$ (cf. formula 1). It implies the current Signal-to-Interference-plus-Noise Ratio $SINR(t)$ and a connectivity forecast $\Delta SINR(t)$ by implementing a connectivity map. After a period of no transmission $(t - t' \leq t_{min})$ continuously a transmission probability $p_T(t)$ as a function of current and future connectivity is computed. In case of success, a transmission proceeds and pCAT returns to the state of no transmission. If no transmission has proceeded until an upper boundary period $(t - t' \leq t_{max})$ is reached a transmission is proceeded immediately and pCAT returns to the state of no transmission.

$$p_T(t) = \begin{cases} 0 & t - t' \leq t_{min}, \\ \left(\frac{SINR(t)}{SINR_{max}}\right)^{\alpha_n \cdot z_1} & t_{min} < t - t' < t_{max}, \ \Delta SINR(t) > 0, \\ \left(\frac{SINR(t)}{SINR_{max}}\right)^{\alpha_n / z_2} & t_{min} < t - t' < t_{max}, \ \Delta SINR(t) \leq 0, \\ 1 & t - t' \geq t_{max}, \end{cases} \tag{1}$$

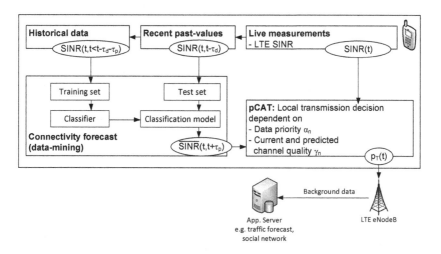

Fig. 1: Scheme of a predictive channel-aware transmission based on data mining

Details regarding this formula can be found in [9]. To avoid mentioned disadvantages of a connectivity map, a location-independent connectivity forecast based on data mining is implemented for a predictive channel-aware transmission (pCAT-DM). Methods and algorithms of data mining lead to a connectivity forecast based on the temporal signal course of a connectivity indicator. It is assumed that the temporal course of a connectivity indicator, like the temporal course of spatial coordinates, has a unique characteristic representing the usage of a specific route with regular, recurring, and space-constrained mobility behavior. By replacing a spacial with a temporal dependency, local variations in a temporal signal course through traffic influences can occur. However, the global trend represents the characteristic of a specific route to predict a future course with a classification method.

When applying recent past-values on the trained classification model, the mean connectivity for a future time frame τ_p is classified. This connectivity forecast based on data mining feeds the predictive component of a transmission probability in pCAT-DM (cf. Fig. 1). A data mining process extracts knowledge from the signal course of a recent past time frame τ_d described by representative attributes (data reduction). Dependent on decision thresholds, a k-Nearest-Neighbor (kNN) classifier predicts a mean future connectivity represented by a label in time intervals of one second. In consideration of prerequisite knowledge about any possible state of the connectivity in a represented system, e.g. influence of pathloss, shadowing, noise and interference, attributes such as current, average (mu), variance (var), min, max and gradient (mReg) values are selected (cf. Fig. 2). In time intervals of one second descriptive attributes are combined with a predictive label and form a data mining example. As a label represents a future change of connectivity, the computation takes a logarithmic quantization

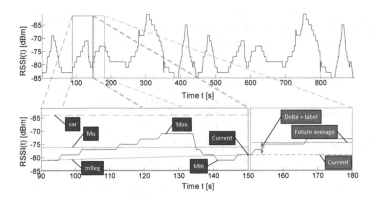

Fig. 2: Attributes within a descriptive frame of $\tau_d = 60s$ and label within a predictive frame of $\tau_p = 30s$ referred to a current time $t = 150s$

into account to converge a uniform distribution of labels, as small future changes appear exponentially more often than large changes.

4   Methods for Performance Evaluation

To evaluate the performance of a connectivity forecast, experimentally measured data represented by a Received-Signal-Strength-Indicator (RSSI) in UMTS technology and data from a simulative reference scenario represented by the SINR in LTE technology are considered (cf. Fig. 3). After one specific route each is used four times to generate data mining examples for a training set, a fifth use of the same route reveals data mining examples that are applied as a test set to a classification model. Whereas experimental data are filtered from a massive data base (cf. [10]) dependent on the spatial focus of a respective route, the simulated data is generated by fusing simulated individual mobility in SUMO and a simulated mobile communication system in LTE-Sim [11][12]. By this a realistic traffic model based on OpenStreetMap- (OSM) street network describes mobility of individual users in a faithful reproduction of an LTE network with a realistic positioning of base-stations, scheduling and channel model. Based on continuous SINR signal curves training- and test-examples are generated. Knowledge about valid labels in a training set is used for a statistical evaluation of the classifier's performance.

  By applying test data to a classification model, prediction labels are classified, which can be repatriated to LTE-Sim to evaluate a gain of pCAT-DM towards periodic transmission on connectivity at transmission and delay between transmissions regarding a mobile device and a reduction of a mean utilization regarding the mobile communication system (cf. Fig. 3). Fig. 4 shows an example route that leads through a section of the simulated reference scenario and four contin-

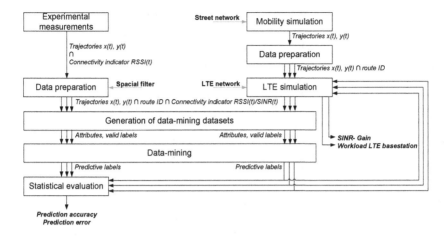

Fig. 3: Structural scheme with regard to methods and tooling for performance evaluation

uous SINR signal curves with a characteristic global trend and local variations to generate a training set.

5   Results

To turn out the advantage of using a kNN classifier, its performance is evaluated in comparison to a pure random process and a Bayesian estimator with an a-priori probability based on the distribution of labels in a traning set. Fig. 5 shows

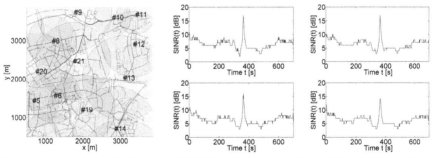

(a) example route in the simu-   (b) SINR signal curves of four different usages of one
lated reference scenario   example route

Fig. 4: Generation of SINR signal curves using an example route as raw data for a data mining process

the comparison of these classifiers dependent on the number of quantization steps of the label (SINR difference between current value and average value in the future with τ_p look ahead window). It can be seen from the figure that kNN reaches an up to four times higher prediction accuracy and a 10 dB lower prediction error in a 95% confidence interval compared to the named reference classifiers. As time-dependent connectivity forecasts are repatriated to LTE-Sim, a transmission probability taking a predictive and channel-aware component into account is computed in time steps of one second. By simulating 200 User Equipments (UE) in an LTE reference scenario, pCAT-DM turns out an immense SINR gain of 11 dB towards periodic transmission (cf. Fig. 6a, 6b). Regarding the delay of transmission time slots of 200 UEs, pCAT-DM reaches the same mean delay of periodic transmission whereas the upper quartile scatters more due to postponing transmission at increasing future connectivity (cf. Fig. 6c). On the side of a mobile radio system, the utilization can be reduced by 33 % with pCAT-DM towards a peridic transmission of the same ammount of data (cf. Fig. 6d).

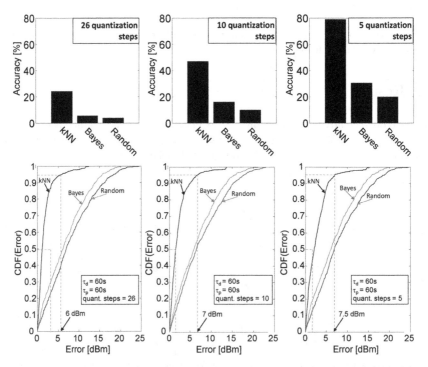

Fig. 5: Classification accuracy as a function of the number of quantization steps, whereas kNN classifier multiplies accuracy compared to Bayesian estimator and pure random process at comparatively low prediction error based on experimental data in a real UMTS scenario

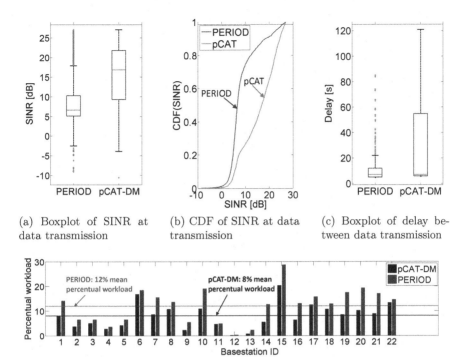

(a) Boxplot of SINR at data transmission

(b) CDF of SINR at data transmission

(c) Boxplot of delay between data transmission

(d) Mean utilization of a mobile communication system reduced by 33 % in pCAT-DM

Fig. 6: Comparison of SINR, delay and utilization of a mobile communication system between pCAT-DM and periodic transmission in a simulative LTE reference scenario

6   Conclusion

In this paper, the time-variant channel quality of cellular communication systems is predicated based on machine learning methods (kNN classifier). The predicted channel quality is leveraged by the pCAT-DM scheme in order to improve data transmission decisions for non-time-critical CPS applications. Main contributions of this paper are:

- Development, implementation and validation of a connectivity forecast based on machine learning methods: prediction of future connectivity based on characteristic signal curves of continuous, historical connectivity information.
- The prediction accuracy mainly depends on the number of quantization steps.
- Using kNN classifier, the prediction error can be reduced by 10 dB compared to a random prediction.
- pCAT-DM, which uses the prediction of the channel quality, improves the SINR for data transmissions significantly and reduces the mean utilization

of a mobile communication system by 33 % in comparison to a periodic transmission.

Acknowledgement

Part of the work on this paper has been supported by Deutsche Forschungsgemeinschaft (DFG) within the Collaborative Research Center SFB 876 Providing Information by Resource-Constrained Analysis, project B4.

References

1. Song, C., Qu, Z., Blumm, N. and Barabsi, A.: Limits of Predictability in Human Mobility. In: Science 327, 2010.
2. Xiao, X., Li, Y. and Kui, X.: Location Patterns and Predictability of Large Scale Urban Vehicular Mobility. In: IEEE Wireless Communications and Networking Conference (WCNC), Istanbul, Turkey, 2014.
3. Michaelis, S. and Wietfeld, C.: Comparison of User Mobility Pattern Prediction Algorithms to increase Handover Trigger Accuracy. In: IEEE 63rd Vehicular Technology Conference, Melbourne, Australia, 2014.
4. Pogel, T. and Wolf, L.: Prediction of 3G Network Characteristics for Adaptive Vehicular Connectivity Maps (Poster). In: IEEE Vehicular Networking Conference (VNC), Seoul, Korea, 2012.
5. Nicholson, A. J. and Noble, B. D.: BreadCrumbs: Forecasting Mobile Connectivity. In: ACM International Conference on Mobile Computing and Networking, New York, USA, 2008.
6. Ide, C., Dusza, B. and Wietfeld,C.: Performance of Channel-Aware M2M Communications based on LTE Network Measurements. In: IEEE Int. Symp. on Personal, Indoor and Mobile Radio Commun., London, UK, Sep. 2013.
7. Ide, C., Dusza, B., Putzke, M. and Wietfeld, C.: Channel Sensitive Transmission Scheme for V2I-based Floating Car Data Collection via LTE. In: IEEE Int. Conf. on Commun., Ottawa, Canada, Jun. 2012.
8. Ide, C., Dusza, B., and Wietfeld, C.: Client-based Control of the Interdependence between LTE MTC and Human Data Traffic in Vehicular Environments. In: IEEE Transactions on Vehicular Technologies, vol. 64, no.5, 2015.
9. Wietfeld, C., Ide, C. and Dusza, B.: Resource Efficient Mobile Communications for Crowd-Sensing. In: 51st ACM/EDAC/IEEE Design Automation Conference (DAC), San Fransisco, USA, 2014.
10. Lewandowski, C., Groening, S. and Wietfeld, C.: A System for Analyzing and Optimizing Urban Electric Vehicle Fleets. In: International Conference on Connected Vehicles and Expo (ICCVE 2012), Beijing, China, 2012
11. Krajzewicz, D., Erdmann, J., Behrisch, M. and Bieke, L.: Recent Development and Applications of SUMO - Simulation of Urban MObility. In: International Journal on Advances in Systems and Measurements, 2012.
12. Piro, G., Grieco, L., Boggia, G., Capozzi, F., Camarda, P.: Simulating LTE Cellular Systems: An Open-Source Framework. In: IEEE Transactions on Vehicular Technology, 2011.

Towards Optimized Machine Operations by Cloud Integrated Condition Estimation

Brecher, C.; Obdenbusch, M; Herfs, W.

Laboratory for Machine Tools and Production Engineering (WZL) of
RWTH Aachen University, Steinbachstraße 19, 52074 Aachen
{C.Brecher, M.Obdenbusch, W.Herfs}@wzl.rwth-aachen.de

Abstract. The requirements concerning the Overall Equipment Effectiveness (OEE) – especially machine availability – increase constantly in production nowadays. Unplanned down-times have to be prevented by efficient methods. Predictive, condition based maintenance represents a valuable approach for fulfilling these demands. Existing concepts lack of information, training data or interconnectedness. The objective of this paper is to present a novel approach in the context of Industrie 4.0 by using machine models with integrated uncertainties in the beginning, resolving these by methods of machine learning during operation and integrating both into a cloud-based service architecture.

Keywords: condition monitoring, machine learning, state estimation, cloud

1 Introduction

Nowadays production systems are subject to significant complexity with high demands for availability, efficiency or costs [1]. Especially machine availability is a strict requirement since it is strongly connected to increasing the added value. Although condition monitoring together with predictive maintenance represents a good approach in order to avoid unplanned down-times, process interruptions due to wear do not lead to component failure in the first place. In fact the complexity of the overall system causes different interruptions like inaccurate positioning of mechanical parts or diminished force transmission.

As the basis for condition monitoring components with risk of default like bearings, belts or drives are continuously monitored, often in an sporadic and manual way [2]. By using powerful algorithms multistage warnings and down-times can be derived for condition-based maintenance, which results in lower economic damages than unplanned failures (secondary damage, loss of production).

In the context of Industrie 4.0 intelligent maintenance is one of the encouraged use cases [3]. In order to realize flexible, dynamic and self-optimizing production Cyber-physical systems are the indispensable basis for building Cyber-physical production systems (CPPS) [4; 5]. At the same time the adaption of existing technologies like cloud computing for enabling local CPPS with global intelligence is absolutely necessary [6; 7]. Furthermore cloud computing is named as one of the key technologies for

adfa, p. 1, 2011.

software-defined platforms and service platforms [8] and foreseen as a disruptive technology for present manufacturing [9].

Putting all these facets of Industrie 4.0, cloud computing, Internet of Things, Internet of Services together the following paper presents a novel approach for cloud integrated condition estimation for optimized machine operations. It begins with a short overview on existing condition monitoring solutions and methods for intelligent state estimation and describes some existing approaches concerning cloud-based reference-architectures. Afterwards state estimation of a belt in a tubular bag machine as a use case for applying methods of machine learning is focused. While wear of components cannot be sufficiently modeled during engineering phase of the system, estimation during operation leads to resolved uncertainties in the machine model. Finally an approach for integrating algorithms and models in a cloud architecture to archive intelligent maintenance is discussed.

2   State of the Art

Up-to-date condition monitoring solutions mainly use online data measurement and aggregation for observing dynamic processes. Usually both internal data from controls (current, torque, speed, e.g.) and external sources (vibration, temperature, e.g.) are considered. By now different approaches and products exist, which will be provided as an overview in the following.

The high grade of technology in production machines and the used measurement and analysis hardware for condition monitoring result in extensive additional costs. Therefore many of the existing solutions address large systems like (wind) power plants, mining or rolling mills where down-times lead to enormous economic losses.

However availability of smaller production systems is increasingly important raising the need for monitoring as well. Usual methods for determining a plant's state are vibration analysis, thermography, vibro-acoustic approaches or oil analysis [10]. Commercial applications often integrate necessary sensors as well as elaborate algorithms. Examples for such systems are "Ω-Guard" by Bachmann, "@ptitude Asset Management System" by SKF, "Real Time Maintenance" by ifm or "SIPLUS CMS" by Siemens.

The majority of commercial products for condition monitoring tasks are applied locally for specific tasks without connection to superior systems and require some time for self-parametrization (learning phase) or configuration by any user.

Many approaches for optimizing condition monitoring are state of research. [11], for example, introduces local monitoring methods for primary components of machine tools like spindles, ball screws and linear guiding. [2] focuses on packaging machines. After the identification of standard wear parts methods and strategies for predictive maintenance by frequency analysis (Peak to Peak, RMS-FFT, frequency selective FFT) are developed. The results are transferred into a platform, which constitutes an exchange of data and expert knowledge above companies' boundaries.

Intelligent machine learning methods for predictive analytics are used in different applications like monitoring drilling processes in oil and gas industry [12] or identifying integral type faults in semiconductor production [13]. [14] suggests a predictive analytics framework for manufacturing as a generalization of different underlying machine learning techniques. Many others like [15] concentrate on process monitoring (tool breakage, collision) with these methods.

In fact existing concepts are very sparse and the majority of the given examples provide solutions for non-production respectively manufacturing environments. This might be due to a lack of significant data objects for machine learning training.

As stated the guiding principles of Industrie 4.0 like interconnection across life-cycle phases, machines and companies, self-optimizing cyber-physical production systems and IT-globalization encourage conceptualizing new approaches.

[16], for example, deals with the challenge of data consistency into cloud platforms and takes aspects like protocols, machine connectivity or safety into consideration when launching the idea of a cloud gateway for the shop floor. [17] describes the adaption of a service-oriented architecture to industrial cloud-based CPS. With these approaches as basis first concepts for specific cloud-integrated condition monitoring tasks develop [18–20].

To conclude, none of the existing approaches cover an adaptive, wear considering process parametrization for production technology by methods of machine learning and the application of cloud technologies.

3   Use Case: Tubular Bag Machine

Tubular bag machines are used to package a wide range of goods and consistency (solid, powdery, fluid). Within the process firstly a rectangular film (1) from a roll (2) is formed into a tube (3) via thermal sealing. This action is supported via two belts (4) which pull off the film. At the same time a horizontal sealing unit (5) forms the bottom and – after the product filling (6) – the top of the bag (7). As presented in **Fig. 1** several forces occur which in fact can lead to process disturbances.

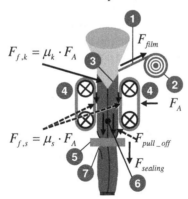

Fig. 1. Occurring forces during pull-off of film in a tubular bag machine

With the speed of the machine (up to 180 cycles per minute) their influence increases. One major use case consists of the interaction between the belts, the film and the forming tube (1, 3, 4). If the contact pressure F_A is incorrectly set a slip of the film will be the result, which directly leads to asynchronous motion and therefore an interruption of the process. Causes for an imprecise parameter setting are a lack of knowledge concerning wear of the belt, unknown film specific parameters (friction coefficient, tensile strength or elasticity, e.g.) or wrong assumed forces for the general pull off process.

4   Approach of an Intelligent Condition Monitoring

In order to optimize existing concepts and dispel existing deficits an approach for intelligent condition monitoring is presented. It mainly consists of four steps:

1. Building rough machine, process or component models during engineering phase [21]
2. Identifying wear induced uncertainties in the models and implement interfaces for external parameter input
3. Conceptualizing an external algorithmic method for state estimation on the basis of current machine sensor data
4. Integrate models and algorithms in a cloud for cooperative modelling and usage of life-cycle over lapping information

4.1   Systems Engineering and Resulting Models

During the engineering phase models like shown in **Fig. 2** are built for simulation purposes or supporting programming efforts, for example. Firstly specific targets for the model based approach with respect to the use case described in chapter 3 have to be defined. The overall aim is to simulate the behavior determining the dependency between contact pressure, torque and static friction (belt-film) as well as the relation between static friction (belt-film), film roll and dynamic friction (film-forming tube). Additionally the drive force F_M should not exceed static friction in order to avoid slip effects:

$$F_M \le F_{f,s} = \mu_s \cdot F_A \tag{1}$$

Since the motor force results from its torque and the radius it can be seen that:

$$M = F_M \times r \Rightarrow F_M = \frac{M}{r} \Rightarrow \frac{M}{r} \le \mu_s \cdot F_A$$
$$\Rightarrow \frac{M}{F_A} \le \mu_s \cdot r = const. \tag{2}$$

That means that if the torque (by current and motor model) and the specific, wear dependent friction was known, the contact pressure could be dimensioned correctly.

These correlations are integrated into a Modelica model (see **Fig. 2**) combined with a component based force transmission of the motor (gearbox, cardan shaft, pinion …) and the pneumatics.

The major uncertainty within this system is the friction coefficient, which changes with the wear of the belt and the film attributes.

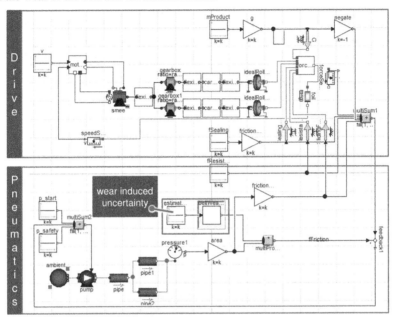

Fig. 2. Modelica model describing the machine components and occurring forces involved in the sealing process

4.2   State Estimation

Like described in [22; 10] oscillations respectively frequency analysis can be used as an indicator for wear of different components (bearing, belt, chain …). In the given use case basic oscillations arising from the tooth engagement at the pinion are modulated by the film take-off unit. If a faulty state of operation is reached changes of the amplitudes within the spectrum can be identified. They can be detected with acceleration sensors (measuring different dimensions) mounted to the take-off unit. As determined before the great challenge is to find an intelligent integrative representation of the different conditions (high-dimensional vectors) since many different spectra in changing states of operation (mainly cycle times) have to be included (see **Fig. 4**).

It was decided to choose the approach of machine learning to satisfy these requirements. In general these methods can be divided into supervised (training set is labelled) and unsupervised algorithms (training set is not labelled). In a first concept the unsupervised clustering by k-means was applied in order to evaluate the procedure with the labelled test data later on. k-means in general is an iterative algorithm. It starts with k random means. Next the data objects are assigned to different cluster centers. Since the target of the algorithm is to minimize the sum over all Euclidian distances (3) respectively the variances of each cluster, each data point is assigned to its nearest cluster. In the final step the centers of each cluster are recalculated by building the mean again. The algorithm stops when no object changes clusters. [23]

Fig. 4 shows the result of the k-means approach reduced to two dimensions for presentation. With the specification of two clusters (new belt and component at the end of its lifetime) the data objects could be separated clearly. With taking the label information into consideration again an error of $E = 0.041$ could be calculated.

$$J = \sum_{i=1}^{k} \sum_{\mathbf{x}_j \in C_i} \| \mathbf{x}_j - \mathbf{\mu}_i \|^2 \qquad (3)$$

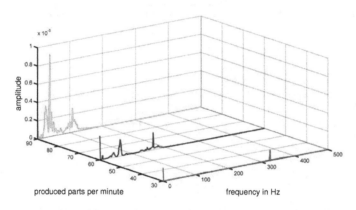

Fig. 3. FFT analysis (spectrum) of measured accelerations at different cycle times (30, 60, 90 cycles/minute)

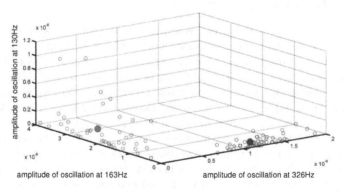

Fig. 4. Clustering by the application of k-means

In a second approach the supervised maximum likelihood estimate was applied. For the given data sets different probability density – Gaussian distribution assumed in specific – were calculated. For the training set a perfect fitting could be archived.

5   Cloud Integration

The primary deficits of existing solutions are the lack of flexibility and overall information, for example for building models, or knowledge exchange. If a new belt is inserted there is no automatism for receiving specific attributes combined with a method for preventive condition monitoring; the same applies for different films.

The introduction of cloud technology and therefore the creation (on the basis of collected data across machines) and later integration of robust models (machines, components, and processes), algorithms and methods in a superior system promise great enhancements. The centralization of data additionally leads to improved models since suppliers gain information from their users. Companies selling packaging materials can provide specific information about their products and data from the shop floor like measured oscillation can be analyzed externally in a cloud architecture for condition prediction. During realization common advantages like performant services (SaaS and PaaS), scalability or pay-per-use payment models can be established.

Fig. 5 shows an extract of the new cloud based reference architecture for condition monitoring currently developed. Machines are connected to the cloud via a middleware since the necessary semantic information is usually not provided sufficiently. Therefore the middleware, for example, augments existing OPC DA interfaces with an extensive OPC UA information model. With this representation a connection with the cloud for production (*Pro*CLOUD) can be established via REST [24].

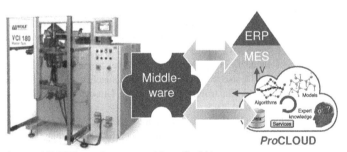

image: Wolff Verpackungsmaschinen GmbH

Fig. 5. Extract of a novel cloud based reference architecture

In the described use case the acceleration data is send to the cloud for analysis purposes. A cloud service consults the machine learning algorithm for determining the current condition of the film take-off belts. This information can be used for predictive maintenance. For an optimized sealing process through calculating the correct contact pressure another service can be called. This one takes the current state of the belt and the persistent machine model (see **Fig. 2**) into consideration.

6   Conclusion and Outlook

The paper presented a novel approach for combining methods of machine learning (clustering by k-means and maximum likelihood estimate) with cloud based services to improve state estimation of mechanical components within a packaging machine in the context of Industrie 4.0. In order to minimize error or to prevent and validate against overfitting the amount of training data (especially different states) can be increased significantly. Other next steps will be the evaluation of different machine learning algorithms and the differentiation of necessary services for maintenance provided by *Pro*CLOUD.

7   Acknowledgement

The authors would like to thank the German Research Foundation (DFG) for the support of the depicted research within the Cluster of Excellence "Integrative Production Technology for High-Wage Countries".

8   References

1. Appelrath, H.-J.; Kagermann, H.; Krcmar, H.: Future Business Clouds. Cloud Computing am Standort Deutschland zwischen Anforderungen, nationalen Aktivitäten und internationalem Wettbewerb. acatech - Deutsche Akademie der Technikwissenschaften e.V., München, 2014.
2. Atmosudiro, A.; Faller, M.; Verl, A.: Durchgängige Datenintegration in die Cloud. Ein Konzept zur cloudbasierten Erfassung von Produktionsdaten. In: wt Werkstattstechnik online. 104. Jg., 2014, Nr. 3.
3. Bechhoefer, E.; Morton, B.: Condition monitoring architecture. To reduce total cost of ownership. In: 2012 IEEE Conference on Prognostics and Health Management (PHM), Denver, Co, 18.-21. June, 2012, Piscataway, NJ, 2012.
4. BITKOM Bundesverband Informationswirtschaft, Telekommunikation und Neue Medien e. V.; VDMA Verband Deutscher Maschinen- und Anlagenbauer e. V.; ZVEI Zentralverband Elektrotechnik- und Elektronikindustrie e. V: Umsetzungsstrategie Industrie 4.0. Plattform Industrie 4.0, April 2015.
5. Brecher, C.; Pohlmann, G.; Herfs, W.: Zustandsbasierte Diagnose an Rollenketten von Verpackungsmaschinen. Höchste Beanspruchung bei 140 Takten/min. In: wt Werkstatttechnik online. 99. Jg., 2009, 7/8.
6. Brecher, C.: Zuverlässige Produktionsanlage (ZuPro). Verbundprojekt im Rahmenkonzept "Forschung für die Produktion von morgen" des Bundesministeriums für Bildung und Forschung (BMBF). 1. Aufl., Aachen: Apprimus-Verl, 2010.
7. Brecher, C. (Hrsg.). ZuPack. Zustandsorientierte Instandhaltung von Verpackungsmaschinen, Aachen: Apprimus-Verlag, 2010.
8. Colombo, A. W.; Bangemann, Thomas, Karnouskos, Statmatis; Delsing, J.; Stluka, P.; Harrison, R.; James, F.; Lastra, J. L. (Hrsg.). Industrial Cloud-Based Cyber-Physical Systems. The IMC-AESOP approach, Switzerland: Springer International Publishing, 2014.
9. Duda, R. O.; Hart, P. E.; Stork, D. G.: Pattern classification. 2. Aufl., New York: Wiley, 2001.

10. Eickmeyer, J.; Pethig, F.; Schriegel, S.; Niggemann, O.; Givechi, O.; Li, P.; Krüger, T.; Frischkorn, A.; Hoppe, T.: Intelligente Zustandsüberwachung von Windenergieanlagen als Cloud-Service. In: Automation 2015. 16. Branchentreff der Mess- und Automatisierungstechnik: Benefits of Change - the Future of Automation: Kongresshaus Baden-Baden, 11. und 12. Juni 2014. Düsseldorf: VDI Verlag GmbH, 2015.

11. Epple, U.: Begriffliche Grundlagen der leittechnischen Modellwelt. Teil 2: Anlagen, Komponenten, Funktionen und Co. In: automatisieren!, 2009, Nr. 12.

12. Fielding, T. R.: Architectural Styles and the Design of Network-based Software Architectures, 2000.

13. Kagermann, H.; Wahlster, W.; Helbig, J.: Umsetzungsempfehlungen für das Zukunftsprojekt Industrie 4.0. Abschlussbericht des Arbeitskreises Industrie 4.0. acatech - Deutsche Akademie der Technikwissenschaften e.V., April 2013.

14. Kagermann, H.; Riemensperger, F.; Hoke, D.; Helbig, J.; Stocksmeier, D.; Wahlster, W.; Scheer, A.-W.; Schweer, D.: Smart Service Welt. Umsetzungsempfehlungen für das Zukunftsprojekt Internetbasierte Dienste für die Wirtschaft. acatech - Deutsche Akademie der Technikwissenschaften e.V., Berlin, März 2015.

15. Kejela, G.; Esteves, R. M.; Rong, C.: Predictive Analytics of Sensor Data Using Distributed Machine Learning Techniques. In: 2014 IEEE 6th International Conference on Cloud Computing Technology and Science (CloudCom). Singapore, 15.-18. December, 2014, Piscataway, NJ: IEEE, 2014.

16. Kolerus, J.; Wassermann, J.: Zustandsüberwachung von Maschinen. Das Lehr- und Arbeitsbuch für den Praktiker. 6. Aufl., Renningen: expert-verlag, 2014.

17. Kuhn, A.: Zukunft der Instandhaltung, Dortmund, 22.03.2013.

18. Lechevalier, D.; Narayanan, A.; Rachuri, S.: Towards a domain-specific framework for predictive analytics in manufacturing. In: 2014 IEEE International Conference on Big Data (Big Data). Washington, DC, USA, 18.-22. August, 2014, Piscataway, NJ: IEEE, 2014.

19. Liang, B.; Hickinbotham, S.; Mcavoy, J.; Austin, J.: Condition Monitoring Under the Cloud. In: Digital Research. Oxford, 2012.

20. VDI/VDE-Gesellschaft Mess- und Automatisierungstechnik; ZVEI Zentralverband Elektrotechnik- und Elektronikindustrie e. V: Referenzarchitekturmodell Industrie 4.0 (RAMI4.0), Düsseldorf, April 2015.

21. Reinhart, G.; Engelhardt, P.; Geiger, F.; Philipp, T. R.; Wahlster, W.; Zühlke, D.; Schlick, J.; Becker, Tilman, Löckelt, Markus; Pirvu, B.; Stephan, P.; Hodek, S.; Scholz-Reiter, B.; Thoben, K.-D.; Gorldt, C.; Hribernik, K. A.; Lappe, D.; Veigt, M.: Cyber-Physische Produktionssysteme. Produktivitäts- und Flexibilitätssteigerung durch die Vernetzung intelligenter Systeme in der Fabrik. In: wt Werkstattstechnik online. 103. Jg., 2013, Nr. 2.

22. Susto, G. A.; Schirru, A.; Pampuri, S.; McLoone, S.; Beghi, A.: Machine Learning for Predictive Maintenance: A Multiple Classifier Approach. In: IEEE Transactions on Industrial Informatics. 11. Jg., 2015, Nr. 3.

23. Teti, R.; Jemielniak, K.; O'Donnell, G.; Dornfeld, D.: Advanced monitoring of machining operations. In: CIRP Annals - Manufacturing Technology. 59. Jg., 2010, Nr. 2.

24. VDI/VDE-Gesellschaft Mess- und Automatisierungstechnik: Industrie 4.0. CPS-basierte Automation, Düsseldorf, Juli 2014.

Prognostics Health Management System based on Hybrid Model to Predict Failures of a Planetary Gear Transmission

Adrian Cubillo[1], Suresh Perinpanayagam[1],
Marcos Rodriguez[2], Ignacio Collantes[2], and
Jeroen Vermeulen[2]

[1] IVHM Centre, Cranfield University
- College Road, Cranfield, Bedfordshire, P.C. MK43 0AL, United Kingdom
[2] ATOS - C/ Albasanz 16, P.C. 28037, Madrid, Spain

a.cubillo@cranfield.ac.uk

Abstract. Health condition monitoring has developed over several years. However, in the area of health assessment algorithms, most of the research has focused on data-driven approaches that do not rely on the knowledge of the physics of the system, while physics-based model (PbM) approaches which rely on the understanding of the system and the degradation mechanisms, are more limited and have the potential to provide more robust predictions due to the understanding of the failure mode phenomena. This paper proposes a Physics-based Model approach to detect incipient metal-metal contact and fatigue degradation of a planetary transmission of an aircraft. Both models are integrated in a real-time Prognostics Health Management (PHM) system that calculates the Remaining Useful Life (RUL) of the component. This tool also incorporates the decision-making process that is performed in the aircraft to connect/disconnect the transmission. A theoretical hybrid model that fuses a machine learning approach with the Physics-based approach to obtain a more robust prediction is also proposed.

Keywords: PHM, planetary transmission, metal-metal contact, prognostics, PbM, IVHM, Integrated Vehicle Health Management

1   Introduction

Health condition monitoring has been a relevant area of research in several industries (Manufacturing, Automotive, Aerospace, etc.) due to its potential to reduce maintenance costs. It is an area of research that aims to recognize changes in the condition of a system that are indicative of a fault; which is particularly interesting for a vehicle or a fleet of vehicles because the available resources and operational demand along with the PHM system can be taken into account to determine the optimal maintenance operations depending on the actual health status of the system [1].

Several engineering aspects are relevant to health condition monitoring: new technologies to manufacture innovative sensors, feature extraction techniques to obtain relevant health indicators, diagnostics and prognostics algorithms, the standardization of the architecture, or the communication between the system and the decision component (particularly important in vehicles). The article is focused on prognostics algorithms.

The information that can be obtained from a health condition monitoring system is normally divided in diagnostics and prognostics. Diagnostics refers to the detection of a fault when it has already occur but additional information can be provided as the failure mode or the identification of the subsystem that has failed. Prognostics not only detects the fault, but also provides an estimation of the RUL with a certain degree of confidence [2]. The algorithms for diagnostics and prognostics are divided in[1, 3–5]:

1. Data-Driven Model (DDM) approaches
2. Physics-based Model (PbM) approaches
3. Hybrid model approaches

DDMs are based on statistical and machine learning techniques and do not directly rely on the knowledge of the physics that govern the system or the degradation mechanism [4]. PbMs are comprehensive mathematical models that describe the physics of the system and their effectiveness depend on the accuracy of the models. One important advantage of PbM is that less experimental data is required compared to DDMs [6] and the synergies between the models used for the design and for PHM make them particularly suitable when developed during the design phase [5]. Hybrid models are formed by a combination of the algorithms mentioned before to obtain a more robust health assessment of the system giving a more robust prediction [7]

An innovative Prognostics Health Management (PHM) solution is presented. The PHM solution proposed predicts the Remaining Useful Life (RUL) of two failure modes of the planetary gear transmission of an aircraft: metal-metal contact on the hydrodynamic bearings and gears fatigue due to variable loading conditions.

Section 2 introduces a description of the planetary gear transmission along with the failure modes and the current health condition monitoring system implemented in the aircraft. Section 3 describes the Physics-based Models that represent the phenomena of both degradation mechanisms. Section 4 describes the real-time algorithm that retrieve the information from the models to estimate the RUL of the transmission. Section 5 presents a technique to combine the information from the PbMs and the DDMs to minimize the errors of the models and provide a more robust estimation of the RUL. Finally Section 6 summarizes the results of the paper and the advantages of an hybrid approach over a PbM approach.

2   Background

The case study presented in this paper, the planetary transmission, is responsible for transmitting mechanical power from the turbine to the generator that supplies electricity for the aircraft systems and its function is to guarantee a constant output speed to the generator under variable input speeds. In order to regulate the speed, a hydraulic unit is responsible for turning one of the ring gears to reduce/accelerate the speed of the output shaft (see **Fig. 1**). Two failure modes are studied: the degradation of the hydrodynamic bearings that allow the free rotation of the planetary gears due to metal-metal contact, and the High Cycle Fatigue (HCF) of the planetary gears under variable loading conditions.

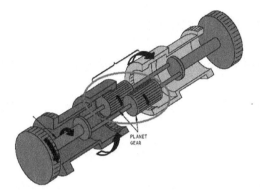

Fig. 1. Schematic differential-marked in red planetary bearings [10]

Metal-metal contact may occur due to insufficient lubrication, excessive loading conditions or increased oil temperature, leading to increased friction and rapid degradation. A prognostics algorithm capable of detecting the degradation in the early stages would minimize the damage to adjacent components.

The fatigue of the gears is caused by excessive loading and it is not monitored in real time, but due to the wide range of operational conditions depending on the flight profile (cruise, landing, taking off or taxi), a health assessment of fatigue life depending on the actual operational conditions instead of a using a standard profile will provide a more accurate prediction of the RUL of the transmission due to fatigue.

3   Physical-based Models

3.1   Metal-Metal Contact Model

The approach to detect metal-metal contact, shown in **Fig. 2**, consists of two coupled models:

1. Stribeck curve model
2. Thermal model

The Stribeck curve model calculates the friction in the bearing while the thermal model calculates the temperature profile along time based on the friction previously calculated; this temperature is used as a health indicator of metal-metal contact. However, both models cannot be decoupled because the Stribeck curve model requires the temperature as an input due to the effect on the viscosity and the thermal model requires the friction as an input to determine the friction heat generated in the bearing.

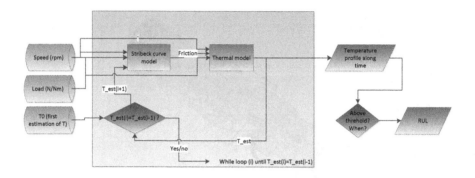

Fig. 2. Metal-metal contact RUL estimation process

The first model, the Stribeck curve model, is used to represent the friction in the hydrodynamic bearings depending on the operational conditions. The model, based on finite differences and developed in Matlab$^{@}$ represents the fully-lubricated region, the elasto-hydrodynamic region for heavy-loaded conditions, and the mixed lubrication region. Metal-metal contact occurs when there is mixed lubrication.

This friction coefficient, obtained as a function of temperature, speed and load from the Stribeck curve model, is used as an input for the thermal model (**Fig. 2**). The thermal model consists of a Finite Elements Analysis (FEA) that executes a quasi-static simulation along time that provides the temperature profile in the transmission (see **Fig. 3**) and updates the friction to consider the actual temperature at every step.

3.2 Gear Fatigue Model

The second failure mode, fatigue of the gears, aims to estimate the RUL of the system for any loading condition by calculating the most critical stresses using a FEA of the teeth of the gear (see **Fig. 4**). The simulation is run for the path

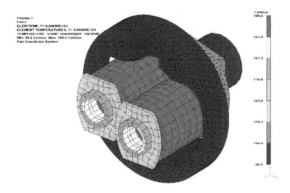

Fig. 3. Thermal model of the transmission (Temperature in Celsius)

in which a tooth is in contact for different loading conditions and a correlation between load and maximum stress is obtained.

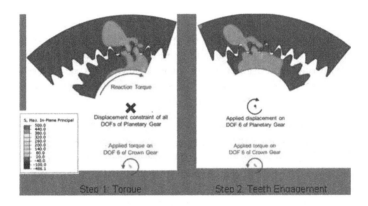

Fig. 4. FEA of planetary gear stresses

A High Cycle Fatigue model, the Basquin law [8], shown in **Eq. 1**, is used to estimate the number of cycles N_f as a function of the maximum stresses amplitude σ_a, where σ_f' is the fatigue strength coefficient and b the fatigue strength exponent.

$$\sigma_a = \frac{\Delta\sigma}{2} = \sigma_f' \left(2N_f\right)^b \tag{1}$$

Due to the variable operational conditions a cumulative damage rule is needed. The well accepted Miners rule [9] is used, as shown in **Eq. 2** where D is the cumulative damage, n_i are the number of cycles under the loading condition i and N_i are the total number of cycles for the given loading condition i. The estimated

RUL due to fatigue damage can be obtained using the real-time loading profile instead of a predefined profile. The total failure is considered when D is above 1.

$$D = \sum_{i=1}^{load_p} = \frac{n_i}{N_i} \qquad (2)$$

4   Prognostics Health Management System

The PbMs described in section 3 are computationally intensive and cannot be executed in real-time on the aircraft. The models should be executed for a wide range of operational conditions on the ground and the PHM system should include a database with the results of the simulations to minimize the number of calculations executed in the aircraft.

In order to implement the information from the models described above and to obtain the RUL additional calculations are needed, which are executed by the PHM system in the aircraft. The PHM system retrieves the data from the models, post-processes them to find the appropriate simulations and computes the RUL of the system. Additionally the system can automatically trigger an alarm and decide to disconnect the transmission. The information provided by the PHM system is reflected on a Graphical User Interface that includes (see **Fig. 5**):

1. RUL estimation (graph on the left - Figure 5)

2. Comparison between the RUL and the flight hours left

3. Temperature estimation along time (graph on the right -Figure 5)

4. Indicators of the status of the transmission and alarms in case any problem is detected.

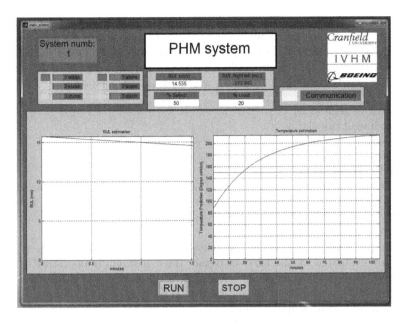

Fig. 5. PHM system graphical user interface

5   Future improvements: Hybrid model approach

The approach presented in the previous sections consists in a theoretical PbM that does not take into account additional data not required by the model and does not consider changes in the environment neither.

In order solve this two limitations two alternatives are proposed for the PbM of metal-metal contact:

1. Parameter estimator to adapt the model to changes in the environment
2. DDM approach based on a time-series ANN
3. Data fusion of PbM with DDM to obtain an hybrid model

5.1   Parameter estimator

PbMs rely entirely on the accuracy of the model, but a model is an approximate representation of the real phenomena, leading to two types of error:

1. Differences between the behaviour of the model and the real phenomena
2. Variability due to environmental factors that are not considered by the model

A parameter estimator is able to minimize those two types of error by comparing the output of the model with a direct or indirect measurement of the health indicator and minimizing the differences. The simplest technique consists in minimizing the mean square value but for non-linear phenomena the Extended Kalman Filter or the particle filter are more suitable. Moreover, the particle filter not only corrects errors of the model, but also provides an estimation of the RUL error deviation, which is particularly important for optimizing the maintenance. Therefore a particle filter is proposed, which consists in 4 phases:

1. Particles generation
2. Likelihood calculation
3. Re-sampling and propagation

Particle Generation:

The particle filter generates a set of particles (a particle refers to a state of the system) for the given operational conditions simulating similar states to the real one. This step is only executed once.

Likelihood calculation

The particles initially generated (step 1) or the re-sampled particles (state i) are evaluated by the PbM to obtain the estimated Temperature of each particle. The temperature of each particles is compared to the measured value and the likelihood of each particle is computed.

Re-sampling and propagation

The particles with low likelihood are eliminated and new particles with higher likelihood are generated. Therefore, only the particles that provide an estimated temperature similar to the measured one remain.

For prognostics the particles are propagated into the future by executing the model with the remaining particles and including the variance of the model. The distribution of the particles into the future is used to provide not only the estimated RUL, but also the uncertainty of the prediction.

It should be noted that a particle filter requires to compare the model with a reference value, in this application the output temperature of the oil. Thus, this value should be known, either by direct measurement (temperature sensor near the bearing outlet flow) or indirect measurement by other means.

5.2   Machine learning approach

The model proposed in previous sections to estimate the RUL of the planetary transmission due to metal-metal contact relies on the sensor data that is used by the model. However, additional sensor data can be relevant and complex patterns can be found to estimate the RUL using a data-driven approach based on machine learning.

The machine learning algorithm is required to detect only one failure mode, metal-metal contact. Thus, machine learning algorithms for fault isolation and fault classification are not considered. Additionally, the algorithm should not only be able to predict the future stationary temperature, but also its evolution along time. Therefore, the algorithm should consider past values of the temperature as well.

The selected algorithm is a supervised time-series ANN, a Feedback Neural Networks that can represent the dynamic phenomena and that uses previous estimations of the temperature as inputs to calculate the future values.

The architecture of the network is shown in **Fig. 6**. Using as inputs: torque, speed, inlet pressure, inlet flow, temperature at the sump, and vibration health indicators (root mean value and power spectral density) to obtain the outlet oil temperature of the bearing as output. The training of the network is done in a test rig capable of measuring all those variables for different operational conditions; thus, generating all the necessary training and test data to validate the network.

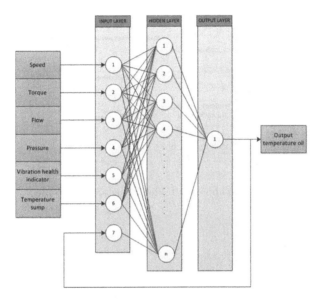

Fig. 6. Feedback Neural Network to detect metal-metal contact

5.3 Data fusion

A particle filter has been proposed to adapt the model to errors and modifications of the environment. However, it requires a reference value of the outlet temperature of the oil that cannot always be available. Additionally an ANN able of predicting the outlet temperature of the oil has been proposed using not only the data used by the PbM, but also information from additional sensors.

An hybrid model that combines both techniques would provide a more robust prediction of the RUL. The simplest approach consists in obtaining a value between the temperature obtained by the ANN and the temperature obtained by the PbM by defining weights for both models as shown in **Eq. 3**; where w_{PbM} refers to the weight of the PbM and w_{ANN} refers to the weight of the ANN, being $w_{PbM} + w_{ANN} = 1$.

$$RUL = w_{PbM}RUL_{PbM} + w_{ANN}RUL_{ANN} \qquad (3)$$

However, if a direct outlet oil temperature measurement is not possible, as it occurs in the current design of the system, both models can be combined by using the estimation of the temperature by the ANN as the reference value for the particle filter that adapts the PbM. This solution does not require the additional sensor to implement the particle filter and combines the advantages of PbM and DDM. The proposed hybrid approach is shown in **Fig. 7**.

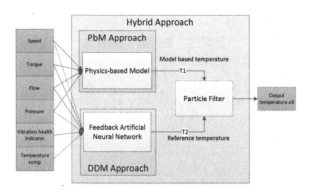

Fig. 7. Hybrid approach to detect metal-metal contact

6   Conclusions

An innovative approach to predict gear fatigue and metal-metal contact in the bearings of a planetary transmission has been presented. The RUL obtained from both failure modes is computed in real-time and is capable of handing variable operational conditions.

It has been shown that complex and computationally expensive models are required in order to accurately represent the physics of the system and its degradation. However, it is necessary to run the models off-line and access a database with the results by the real-time application to obtain acceptable computation times.

The main drawback of a pure PbM approach is the lack of feedback to correct errors in the model and changes in the environment. A particle filter to adjust the model to the real conditions has been proposed for metal-metal contact to solve these two sources of error. However, the direct measurement of the health indicator required by the particle filter is not always available.

The second improvement to the current PbM algorithm of metal-metal contact consists in an Feedback Neural Network to estimate the outlet temperature of the oil. This network uses information from additional sensors and in combination with the PbM approach can provide more reliable results. Two data fusion techniques are considered:

1. A simple rule based on weights of both RUL_{PbM} and RUL_{ANN} to estimate the final RUL. An outlet flow temperature sensor is required to implement the particle filter.
2. The combination of the PbM along with the particle filter using the outlet temperature of the oil estimated by the ANN as the reference value for the model. The outlet flow temperature sensor is not required.

The main advantage of the second option is that a particle filter can be implemented without requiring an additional temperature sensor that measures the output temperature of the oil; thus, without incorporating additional sensors into the real system.

7   Acknowledgements

The research leading to these results has received funding from the European Union Seventh Framework Programme (FP7/2007-2013) under Grant Agreement n° 605779 (project RepAIR). The text reflects the authors' views. The European Commission is not liable for any use that may be made of the information contained therein. For further information see http://www.rep-air.eu/.

References

1. Jennions, I. K. (2011), Integrated vehicle health management: perspectives on an emerging field, SAE International, Warrendale, Pa.
2. Sikorska, J. Z., Hodkiewicz, M. and Ma, L. (2011), Prognostic modelling options for remaining useful life estimation by industry, Mechanical Systems and Signal Processing, vol. 25, no. 5, pp. 1803-1836.
3. Eker, O., Camci, F. and Jennions, I. (2012), "Major Challenges in Prognostics: Study on Benchmarking Prognostics Datasets", Proc.of PHM.
4. Heng, A., Zhang, S., Tan, A. C. C. and Mathew, J. (2009), "Rotating machinery prognostics: State of the art, challenges and opportunities", Mechanical Systems and Signal Processing, vol. 23, no. 3, pp. 724-739.
5. Jianhui Luo, Namburu, M., Pattipati, K., Liu Qiao, Kawamoto, M. and Chigusa, S. (2003), Model-based prognostic techniques [maintenance applications], AUTOTESTCON 2003. IEEE Systems Readiness Technology Conference. Proceedings, pp. 330.
6. Heng, A., Zhang, S., Tan, A. C. C. and Mathew, J. (2009), Rotating machinery prognostics: State of the art, challenges and opportunities, Mechanical Systems and Signal Processing, vol. 23, no. 3, pp. 724-739.
7. Baraldi, P., Mangili, F. and Zio, E. (2012), A Kalman Filter-Based Ensemble Approach With Application to Turbine Creep Prognostics, Reliability, IEEE Transactions on, vol. 61, no. 4, pp. 966-977.

8. Basquin, O. H. (1910) The exponential law of endurance tests, vol. 10 (II), Proceedings of ASTM, pp. 625-630.
9. Miner, M. A. (1945), Cumulative damage in fatigue, vol. 12, no. J. Appl. Mech, pp. 159-164.
10. Makris, K. (2015), http://www.k-makris.gr/AircraftComponents/CSD/C.S.D.htm.

Evaluation of
Model-Based Condition Monitoring Systems
in Industrial Application Cases

S. Windmann[1], J. Eickmeyer[1], F. Jungbluth[1],
J. Badinger[2], and O. Niggemann[1,2]

[1] Fraunhofer Application Center IOSB-INA, Lemgo, Germany,
[2] Institute Industrial IT (inIT), Lemgo, Germany

Abstract. In this paper, model-based condition monitoring methods
are investigated. Reliable process monitoring allows costs and risks to
be reduced by the early detection of faults and problems in the process
behavior and the prevention of component failures or in extreme cases
a production stop of the complete plant. The principal of model-based
condition monitoring consists of comparing the actual process behavior
with the behavior as predicted from process models. For this purpose,
a Hidden Markov Model and a method based on principal component
analysis are applied. Both methods are evaluated in industrial application cases. In doing so, F-measures of 88.25% and 98.84% are achieved
for a wind power station and a glue production plant, respectively.

1 Introduction

Users often fail to get an overview of the current plant status since data is spread
over different computer subsystems or hierarchies. In many cases, it is difficult
for the user to find an anomaly in a large amount of displayed signals, e.g. an un-
expected decrease of the temperature values obtained from a particular sensor.
Such demands lead to an overstraining of an operator's capability to monitor
and diagnose complex systems and renders automatic condition monitoring de-
sirable. In this paper, model-based condition monitoring of strictly continuous
systems is considered. The principle of model-based condition monitoring consists of comparing the actual process behavior with the behavior as predicted
from process models (see Fig. 1, operation phase). The purpose is to detect deviations in the system behavior from the normal state (for example too high or too
low energy consumption of a conveying system). In doing so, significant outliers
can be detected and displayed in a human machine interface. The process models
can be learned from process data collected from the plant and its components
in normal, fault-free operation, as depicted in Fig. 1 (learning phase).

A lot of research has been conducted with respect to the modeling of continuous processes. Clustering-based fault detection methods create groups of strongly
related objects and objects which do not strongly belong to any cluster [1]. Neural networks and regression-based methods have been used to approximate the

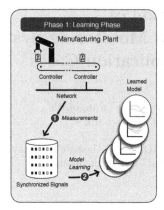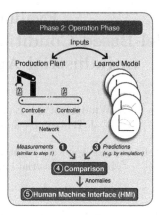

Fig. 1. Model-based condition monitoring.

functional dependency between continuous process variables and time [2]. Sensor signals are predicted according to this functional dependency and significant deviations of predicted signal values from the observations are reported as faults. Stochastic approaches to fault detection are predominantly based on building a probability distribution model and considering how likely objects are under that model [3]. Statistical tests are used to assess the likelihood of faults (see e.g. [4], [5]). In most of the stochastic approaches, state variables are employed for modeling the temporal transitions of hidden process variables, which are related to the measurements with a measurement model. Common approaches are Kalman filters (e.g. [6], [7], [8]), particle filters [9] or Hidden Markov Models (HMMs) [10].

The main contribution of this paper is the evaluation of condition monitoring approaches, which are based on HMMs and principal component analysis (PCA), in industrial application cases. The investigated application cases are the condition monitoring of wind power stations and glue production plants. The remaining part of the paper is structured as follows: Section 2 outlines the investigated model-based condition monitoring approaches. Evaluation is conducted in section 3. Section 4 gives a conclusion.

2   Model-based condition monitoring

In the subsequent section, the investigated HMM-based and PCA-based condition monitoring methods are outlined.

2.1   HMM-based condition monitoring

A convenient representation for piece-wise stationary processes is the Hidden Markov Model (HMM) [11]. This are e.g. processes where energy consumers are switched on and off so that the energy consumption changes stepwise with each switching operation.

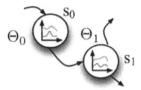

Fig. 2. Hidden Markov Model.

In this approach, no direct dependency between subsequent observations $y(k-1)$ and $y(k)$ at time instances $k-1$ and k is modeled. Instead, discrete system states $s(k)$ are assumed, which correspond to system modes with stationary process behavior. The transitions between discrete system states $s(k-1)$ and $s(k)$ are modeled with a time-invariant matrix of transition probabilities. Furthermore, a relation between observations $y(k)$ and system states $s(k)$ is assumed. In [12], this dependency is modeled with a Gaussian distribution for each system state. Parameter estimation can be accomplished by application of the Baum-Welch algorithm (see [12]). HMMs allow to calculate beside the probability density $p(y(k)|y(0)\dots y(k-1))$ the probability density $p(y(k-l+1)\dots y(k)|y(0)\dots y(k-l))$ of complete blocks of length l. In doing so, the probability density

$$p(y(k-l+1)\dots y(k)y(0)\dots y(k-l)) = \frac{max_{i=0\dots m-1}\theta_i(k)}{max_{j=0\dots m-1}\theta_j(k-l)} \qquad (1)$$

is obtained from the forward variable $\theta_i(k)$, which is in return obtained with the common Viterbi algorithm. Fault detection is accomplished by the evaluation of the probability density $p(y(k-l+1)\dots y(k)|y(0)\dots y(k-l))$ with respect to a given threshold (see [12]).

2.2 PCA-based condition monitoring

In many continuous processes, state transitions can be omitted or cannot be unambiguously characterized. In this case, PCA-based approaches are an alternative, which works on a data base. Each point in this data base represents the complete sensor data at a given time instance. This interpretation leads to a high-dimensional problem, which can be reduced by dimensionality reduction (e.g. by principal component analysis - PCA) [13]. Dimensionality reduction leads to a compact representation of system behavior. In the next step, patterns are extracted, which represent the normal behavior of the system. Fig. 3 shows the data of a wind power station. The area of normal operations is depicted

Fig. 3. Example of dimensionality reduced cluster data.

with green points, while known errors are shown as red points. The interference is small so that the green cluster can be used as model of the normal plant operation. Deviations of the learnt model of normal operation can be related to failures. In [13] the distance $d(y_{pca}(k), x_{pca}(k))$ of PCA-transformed measurements $y_{pca}(k)$ with respect to PCA-reduced data points x_{pca} of the normal behavior has been calculated by application of Marr-Wavelets

$$d(y_{pca}(k), x_{pca}(k)) = \frac{2}{\sqrt{3}\sigma\pi^{\frac{1}{4}}} \left(1 - \frac{\|y_{pca}(k) - x_{pca}(k)\|^2}{\sigma^2} \right) e^{-\frac{\|y_{pca}(k) - x_{pca}(k)\|^2}{2\sigma^2}}$$

$$(2)$$

The standard deviation σ of the Marr-Wavelets is obtained from training data. For $d(y_{pca}(k), x_{pca}(k)) > 0$, a normal process state is assumed at time instance k. Distances $d(y_{pca}(k), x_{pca}(k)) \leq 0$ indicate potential faults, which are displayed to the user.

3   Evaluation in industrial application cases

3.1   Wind power station

PCA-based condition monitoring has been initially evaluated with a data set of a wind power station, which contains data collected in a time span of four years in intervals of 10 minutes (see [13]). Evaluation has been conducted for a time span of 80 days. The remaining data has been used for model learning. In the evaluation phase, 11544 observations in 7013 normal states and 4531 failure states have been analyzed. Condition monitoring results are displayed in table 1. True positives, true negatives, false positives and false negatives are shown in this table. Furthermore, the balanced accuracy and the F-measure, i.e. the harmonic mean of specificity and precision, have been used as quality measures. An overall balanced accuracy of 90.27% and an F-measure of 88.25% have been achieved, respectively. A computation time of 68s was required to evaluate the collected 11544 observations.

3.2   Glue production plant

Further evaluation data was logged in a glue production plant of the Jowat SE, during manufacture of a product. In total, 16 production cycles were logged.

Table 1. Evaluation results for the wind power station

True Pos.	True Neg.	False Pos.	False Neg.	Bal. Acc.	F-Measure	Elapsed Time
3970	6517	496	561	90.27%	88.25%	68s

The modeled part of the system is the input raw material subsystem, which contains six material supply units (smaller containers) connected to a large container where material is mixed. The continuous output variable whose dynamics was learned is the rotation speed per minute of the scraper in the container, which is used to strip the paste from the edge to ensure an even and homogenous mixture. Hidden Markov Models and PCA models have been learned on a set of 13 production cycles. Evaluation has been conducted on three separate production cycles (649382 data points in total). The results are depicted in table 2. The HMM-based condition monitoring (HMM-based CM) yielded a balanced accuracy of 93.11%, while the PCA-based condition monitoring (PCA-based CM) obtained an accuracy of 99.81%. Furthermore, both algorithms have been evaluated with respect to the F-measure. The HMM-based CM achieved an F-meaure of 91.54% while the PCA-based CM obtained an F-measure of 98.84%. PCA-based CM was particularly observed to outperform the HMM-based CM in regions of nonsteady system behavior.

Table 2. Evaluation results for the glue production plant

Method	Bal. Acc.	F-Measure
HMM-based CM	93.11%	91.54%
PCA-based CM	99.81%	98.84%

4 Conclusion

In the present work, HMM-based and PCA-based condition monitoring methods have been evaluated in two industrial application cases. The PCA-based condition monitoring method yielded an overall F-measure of 88.25% for a wind power station. For condition monitoring in a glue production plant, F-measures of 91.54% and 98.84% were obtained by application of the HMM and the PCA, respectively. PCA-based condition monitoring was particularly observed to outperform the HMM-based condition monitoring in regions of nonsteady system behavior.

References

1. C. Frey, "Diagnosis and monitoring of complex industrial processes based on self-organizing maps and watershed transformations," in *Computational Intelligence*

for Measurement Systems and Applications, 2008. CIMSA 2008. 2008 IEEE International Conference on, july 2008, pp. 87 –92.

2. A. Vodenčarević, H. Kleine Buning, O. Niggemann, and A. Maier, "Identifying behavior models for process plants," in *Emerging Technologies & Factory Automation (ETFA), 2011 IEEE 16th Conference on*, 2011, pp. 1–8.

3. A. Shui, W. Chen, P. Zhang, S. Hu, and X. Huang, "Review of fault diagnosis in control systems," in *Control and Decision Conference, 2009. CCDC '09. Chinese*, june 2009, pp. 5324 –5329.

4. S. Windmann, S. Jiao, O. Niggemann, and H. Borcherding, "A Stochastic Method for the Detection of Anomalous Energy Consumption in Hybrid Industrial Systems," in *INDIN*, 2013.

5. S. Windmann and O. Niggemann, "Automatic model separation and application to diagnosis in industrial automation systems," in *IEEE International Conference on Industrial Technology (ICIT 2015)*, 2015.

6. B. C. Williams and M. M. Henry, "Model-based estimation of probabilistic hybrid automata," Tech. Rep., 2002.

7. F. Zhao, X. D. Koutsoukos, H. W. Haussecker, J. Reich, and P. Cheung, "Monitoring and fault diagnosis of hybrid systems," *IEEE Transactions on Systems, Man, and Cybernetics, Part B*, vol. 35, no. 6, pp. 1225–1240, 2005.

8. S. Narasimhan and G. Biswas, "Model-based diagnosis of hybrid systems," *IEEE Transactions on systems, manufacturing and Cybernetics*, vol. 37, pp. 348–361, 2007.

9. M. Wang and R. Dearden, "Detecting and Learning Unknown Fault States in Hybrid Diagnosis," in *Proceedings of the 20th International Workshop on Principles of Diagnosis, DX09*, Stockholm, Sweden, 2009, pp. 19–26.

10. S. Windmann, F. Jungbluth, and O. Niggemann, "A HMM-Based Fault Detection Method for Piecewise Stationary Industrial Processes," in *IEEE International Conference on Emerging Technologies and Factory Automatio (ETFA 2015)*, 2015.

11. L. R. Rabiner, "A tutorial on hidden markov models and selected applications in speech recognition," *Proceedings of the IEEE*, vol. 77, pp. 257–286, 1989.

12. S. Windmann, O. Niggemann, and H. Stichweh, "Energy efficiency optimization by automatic coordination of motor speeds in conveying systems." in *IEEE International Conference on Industrial Technology (ICIT 2015)*, 2015.

13. J. Eickmeyer, P. Li, O. Givehchi, F. Pethig, and O. Niggemann, "Data driven modeling for system-level condition monitoring on wind power plants," in *26th International Workshop on Principles of Diagnosis (DX 2015)*, 2015.

Towards a novel learning assistant for networked automation systems

Yongheng Wang, Michael Weyrich

Institute of Industrial Automation and Software Engineering, University of Stuttgart,
Stuttgart, Germany
yongheng.wang, michael.weyrich@ias.uni-stuttgart.de

Abstract. Due to increasing requirements on functionality (e.g. self-diagnosis, self-optimization) or flexibility (e.g. self-configuration), future automation systems are demanded to be more and more intelligent. Therefore the systems are desired to learn new knowledge from other systems or its environment. The purpose of this work is to propose a prospective concept of learning assistant for networked automation systems. With the help of the assistant, an automation system can obtain new knowledge by collaborating with other systems to improve its prior knowledge. So that the system user is liberated from continuously providing new knowledge to an automation system.

Keywords: learning assistant · networked automation systems

1 Introduction and Motivation

Nowadays, industrial automation systems are always demanded more functionality (e.g. self-diagnosis, self-optimization) or flexibility (e.g. self-configuration). In this case, they are developed to be more and more complex with increasing integration of information and communication technology. The situation becomes even more complex, when the automation systems are networked together. Therefore, industrial automation systems are more difficult to be operated than before. In order to ease the operability of these systems, assistance systems are introduced to support system users, who can then neglect the complexity of the systems.

The intelligence of assistant systems lies in their knowledge base, which has been predefined by the system developer. The knowledge is nothing more than a collection of facts, events, beliefs, and rules, organized for systematic use [1]. There has been a lot of research on intelligent assistants. Most of the concepts have the following two features: (1) the intelligent assistants are always developed for an individual automation system which is not connected to other systems; (2) the prior knowledge of the system will not be changed or only changed in a small scope.

With the emerging of industry 4.0, automation systems are required to connect with each other and coordinate themselves in different responsibilities. In this context, the intelligent assistant should be improved with the consideration of a networking environment and learning new knowledge from other systems in the same network. This art

adfa, p. 1, 2011.

of assistant is named learning assistant. Compared to conventional assistants, the learning assistant is able to adapt to a new or different environment.

The following is the outline of this work. In section 2, a survey on the existing research on assistant system is introduced. In section 3, the requirements on learning assistant to support the learning process in a network environment are discussed. In section 4, the concept of a learning assistant is proposed. A preliminary application scenario is presented in section 5, which is followed by a brief conclusion.

2   State of The Art

Nowadays, intelligent assistants have been utilized in industrial automation systems within a wide application area. A survey of them has been made in this section.

In order to ease the cooperation between human and robot during assembly process, a cognitive system is developed in [2], where the cognitive simulation model (CSM) can be considered to be an assistant. The CSM unit has a prior knowledge and a simulation module to generate an assembly plan and to optimize the assembly sequence. The prior knowledge involves the knowledge of the assembly process, e.g. feasible assembly sequence. It has to be provided by the developer. Obviously the CSM can work well in mass production. However in the view of industry 4.0, which aims to produce individual products, the assembly sequence and parameters of each individual product may vary. Therefore, a prior knowledge is required to be extendable, so that it is able to adapt to the new individual products. A similar example can be found in [3], where a fault prevention model is introduced. The prior knowledge (about system profile, error prevention functionality, abnormity and error, etc.) of fault prevention model determines the performance of the fault prevention. More knowledge means more errors or faults can be recognized. However, in case of large production plants, which involves the net-working of several automation systems, the prior knowledge of each automation system is not enough to predict the error of the whole production plant.

In [4] an ontology based assistant is introduced, which extracts knowledge by a KDD (Knowledge Discovery in Databases) Process Model from large amount of data. However, it is not cleared that, how the existing model can be extended. In the DFG project „Kontext- und Community-basierte Assistenzsysteme für Personen mit Behinderungen (D02)" a context- and community based assistant for disabled people is discussed. The system is used as an intelligent navigation system, which relies on environment model and calculates the route plan automatically. The system can be improved by integrating the communication ability, which allows the system to learn from other systems to adapt more new environment.

There are still more examples of the intelligent assistant. There is no need to mention all of them. Mr. Jasperneite and Niggemann have already concluded the common model of intelligent assistant in [5]. An intelligent assistant (see Fig. 1) always possesses the system objective and a prior knowledge being comprised of domain and equipment knowledge. The system is able to identify its current situation, and then find the solution strategy based on its prior knowledge and system objective.

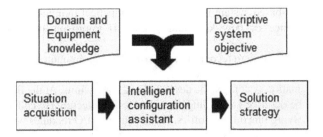

Fig. 1. Conventional intelligent assistant [5]

3   Idea of the Learning Assistant

As concluded in the previous section, the conventional assistant cannot help the systems to obtain new knowledge from its environment. For this reason the idea of a novel learning assistant is introduced as an extension to the conventional assistance (see Fig. 2).

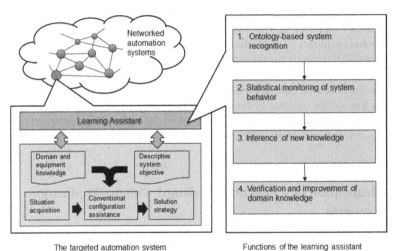

Fig. 2. Idea of learning assistant

In the scenario it is assumed that an industrial automation system is a part of networked systems. The system has a conventional assistant which can determine the solution strategy based on a prior knowledge and predefined system objective. As explained in the previous section, the prior knowledge (domain knowledge) is usually not enough to counter all the possible situations in the networked environment. In this case, the system model is not valid in some situations. For this reason, the proposed learning assistant can search and interact with the similar systems in the network, so that it is able to obtain new knowledge. The intelligent assistance has the following four basic functions: (1) ontology-based system recognition, (2) statistical monitoring of system behavior, (3) inference of new knowledge, and (4) improvement of the domain knowledge.

The prerequisite of system recognition is a compatible communication interface and protocol between systems. Each system will be described by a unique ontology, whereas an automation system can be described in 2 dimensions. At the first dimension the system will be described according to its devices architecture (so called equipment knowledge, involving functional profiles, inputs, outputs, I/O modules, variable types, constraints, etc.). While at the second dimension, the system is described according to its domain knowledge (e.g. diagnosis knowledge). In this respect, the ontology-based device description in [6] and ontological learning assistant in [4] will be employed. The ontology-based system description enables the possibility of communication and accessing between systems.

After step one, the system behavior of all relevant systems in the network will be monitored with respect to the knowledge base. The reason is that, the knowledge base of the targeted system is usually constructed by a limited number of samples, which don't represent the real facts. A typical of this phenomena is the concept drift which has been discussed in [7]. Therefore, the statistical monitoring is able to show which system behavior is strange, and lies away from the typical behavior in the knowledge space. The advantage of step 2 is reduce the human work by means of extending the system behavior model by itself during operation time, instead of considering all behavior possibilities during the development phase. At step 3, the interpretation of these system behavior will then be compared in order to derive new knowledge. If a new interpretation based on other systems can be accepted by the system user, the corresponding knowledge of other systems can be integrated into the target system. When the newly derived knowledge has been verified by an expert, it is demanded to introduce step 4, which will verify and improve the domain knowledge of the targeted system.

4   Realization of a Learning Assistant with an Application Scenario

The previous described idea shows the framework of the learning assistant. The recent section will present how to realize the learning assistant through an application scenario (see Fig. 3). The application scenario can be considered as an extension based on [8] and [9], which have proposed a learning assistant for a single image processing system. For the extension, the application scenario assumes a network with image processing systems in variant countries (e.g. Germany, Spain and Italy). The image processing system is denoted as blue points in Fig. 3. The image processing systems are developed to classify the fruits into different classes. Each of the system has a prior knowledge about the classification of fruits. In order to ease the illustration, it assumes that the systems have learned the different classifications of fruits by using the learning method of Support Vector Machine (SVM). Therefore the prior knowledge is considered to be class boundaries in the feature space. Learning new knowledge means to adapt the class boundaries in the SVM feature space. Based on the assumption, the following content illustrates the functions of the learning assistant in a networked image processing systems.

The first function of the learning assistant is ontology-based recognition of nearby systems. In order to realize this, the individual ontology system has to be developed, including the components, functions, database, knowledgebase, access, security and so on. Since the complexity of creating an ontology for all the networked systems, it is assume that the networked systems can access each other easily.

During the first step, only the systems that have the similar knowledgebase will be remained. This step provides the possibility of learning knowledge from other systems. The second function of the learning assistant is to monitor the classification result of the image systems. The reason can be declared in [9]. The number of training samples for machine learning methods is usually limited, therefore the prior knowledge generated by machine learning is usually not valid during long time operation in the real application. In this case, monitoring of the classification result can help to discover real (fruit) data distribution, which may deviate from the training samples to generate the prior knowledge. As the result of the second function, a stack of points (each point stands for a fruit in the feature space), which are different from the sample points, are identified.

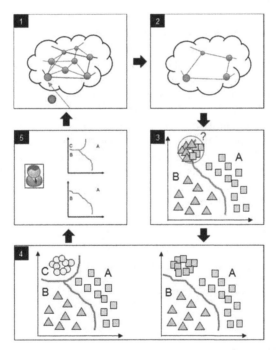

Fig. 3. Application scenario of learning assistant with the example of SVM classifier

The third function of the learning assistant is the inference of new knowledge. The new knowledge is generated by incremental learning, which adapts the old class boundaries to the new found points (being identified at the second step). The prerequisite of using incremental learning is that, the class labels of the new found points should be given. For this purpose, the class labels of the new found points can be generated by two ways: (1) the class labels of the new found points are generated by other systems; (2) the class label of the new found points are generated by the data consistency test being described in [9]. The data consistency test is to analyze the data by using variant distance measurements: Euclidean, Mahalanobis, Bhattacharyya. Based on the measured distance, the class labels of the new found points can be suggested. The result of this function is two new generated classifiers, which have already adapted the class boundaries for the new found points.

The new function of the learning assistant is the verification and improvement of domain knowledge. To approach this, two new generated classifiers will be tested, and the classification accuracies of them will be compared. The user can choose the classifier which has higher classification accuracy. Afterwards the chosen classifier can replace the previous one in the targeted system.

5   Conclusion

The objective of this work is to propose a concept of a learning assistant, which learns new knowledge from other systems in the same network.

Usually a conventional assistant possesses only a certain amount of prior-knowledge, which helps the automation systems adapt to the environment. However prior-knowledge is obtained by only a few number of observations or samples. In this case, the system user always needs to provide the new knowledge to automation systems. Therefore, the recent assistant is not able to adapt to a changing environment automatically. The proposed learning assistant consists of 4 functions: ontology-based system recognition for providing the communication between systems, statistical monitoring of system behavior for detecting unknown system behavior, inference of new knowledge based on the unknown behaviors and evaluation of generated new knowledge. The proposed learning assistant is able to access the knowledge base of other systems, and improve its own knowledge base in order to accommodate itself to changing environment. Especially in the emerging world with internet of things and cyber physical system, the system developer is not able to predefine all the necessary knowledge for a system, but the system is desired to learn new knowledge from other systems in the network.

Reference

1. International Organization for Standardization: Information technology, ISO 2382-28:1997, (2011)
2. Schlick, C.M., et al.: Erweiterung einer kognitiven Architektur zur Unterstützung der Mensch-Roboter-Kooperation in der Montage. 239-263, (2014)

3. Bordasch, M., Göhner P.: Fault Prevention in Industrial Automation Systems by means of a functional model and a hybrid abnormity identification concept. Industrial Electronics Society, IECON 2013-39th Annual Conference of the IEEE, pp. 2845-2850. IEEE (2013)
4. Choinski, M., Chudziak, J.: Ontological learning assistant for knowledge discovery and data mining. In: Computer Science and Information Technology, 2009. IMCSIT'09. International Multiconference on pp. 147-155. IEEE (2009)
5. Jürgen, J.; Niggermann, O.: Systemkomplexität in der Automation beherrschen. atp edition-Automatisierungstechnische Praxis 54.09, 36-45. (2012)
6. Dibowski, H., Kabitzsch, K.: Ontology-based device descriptions and triple store based device repository for automation devices. In Emerging Technologies and Factory Automation (ETFA), 2010 IEEE Conference, pp. 1-9. IEEE (2010)
7. Zliobaite, I.: Learning under Concept Drift: an Overview, Technical Report, Faculty of Mathematics and Informatics. Vilnius University (2009)
8. Wang, Y., Weyrich, M.: An adaptive image processing system based on incremental learning for industrial applications. In Emerging Technology and Factory Automation (ETFA), 2014 IEEE, pp. 1-4. IEEE (2014)
9. Wang, Y., Weyrich, M.: An assistant for an incremental learning based image processing system. In Industrial Technology (ICIT), 2015 IEEE International Conference, pp. 1624-1629. IEEE (2015)

Efficient Image Processing System for an Industrial Machine Learning Task

Kristijan Vukovic[1], Kristina Simonis[2], Helene Dörksen[1], and Volker Lohweg[1]

[1] inIT - Institute Industrial IT, Ostwestfalen-Lippe University of Applied Sciences,
D-32657 Lemgo, Germany
[2] ITA - Institute for Textile Technology, RWTH Aachen, D-52074 Aachen, Germany

Abstract. We present the concept of a perceptive motor in terms of a cyber-physical system (CPS). A model application monitoring a knitting process was developed, where the take-off of the produced fabric is controlled by an electric motor. The idea is to equip a synchronous motor with a smart camera and appropriate image processing hard- and software components. Subsequently, the characteristics of knitted fabric are analysed by machine-learning (ML) methods. Our concept includes motor-current analysis and image processing. The aim is to implement an assistance system for the industrial large circular knitting process. An assistance system will help to shorten the retrofitting process. The concept is based on a low cost hardware approach for a smart camera, and stems from the recent development of image processing applications for mobile devices [1–4].

1 Introduction

Knitted fabrics are composed of one or more threads or of one or more thread systems creating a textile fabric by loop formation. Knitted fabrics can be subdivided into weft and warp knitted fabrics [5]. Weft knitted fabrics are produced by individually moving needles forming stitches successively, whereas in warp knitting all the needles are moved together synchronously. There are basically three different structural elements that make up the weft knitting patterns: knit, tuck and miss (see Fig. 1).

The knitting process has to be set up after changing either yarn, parameters or patterns. There are four important set-up parameters for the knitting process: yarn tension, sinking depth (mesh height), production speed as well as the tension to the fabric controlled by the take-down motor. Investigations have shown that the setting of the take-down rollers driven by an electronic motor affects the length of the loops directly. The knitting process has to be adjusted manually by an operator. In doing so, the operator sets up the parameters according to his mostly longtime experience. In times of demographic change and skills shortage the defects as well as the machine set-up process in case of material or pattern change is to be systemised.

A new approach at inIT and ITA is to combine the assisting system and Machine Learning concept with the knitting process in order to save time and

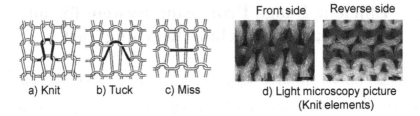

Fig. 1. (a)-(c) weft knitting elements [6]. Light microscopy picture of weft knitted fabric - just knit elements (d).

costs in the knitting retrofitting process of circular knitting machines. The idea is to equip a synchronous motor with sensors, aiming at realising a perceptive motor, which can *feel* and *see*, in terms of a cyber-physical system (CPS).

The theoretical background for a *feeling motor* is described in [7]. That work shows that it is possible to use a motor as a sensor while it is running and doing its usual work. Due to its parallel function as a sensor and actor, the motor can be regarded as an entity that can feel what it is doing. The approach is based on the analysis of the motor currents. It was originally developed for machine-health monitoring to detect the beginning of damages in the bearings at the shaft of the motor before it harms the motor and the devices linked to it. The approach can be adapted to other tasks than health monitoring, e.g. feeling the force at the shaft of the take-down motor in a knitting-machine to draw conclusions about the expansion of the material. Further developments of ML approaches for machine-health monitoring are presented in [8, 9]. It investigates monitoring in Big Data environments [8] and automated performance of machine-health diagnosis [9]. We expect the techniques presented in [7–9] to be suitable for the monitoring of the *feeling motor* in the knitting process.

In the frame of this paper, we concentrate on the realisation of the *vision part*. In the context of that, image processing methods are required. This paper shows an approach for the recognition of different fabrics and a possibility to measure its expansion during the production process via efficient image processing methods. Machine-Learning methods will be needed in order to realise such a system. The features of all possible produced fabric types need to be learned. This includes the feature extraction for the classification of the fabric as well as the measurement of the expansion. This work shows a practical image processing application as part of a CPS for a knitting process. Because high costs are a barrier for image processing systems in the knitting industry, this approach aims at low complexity to ensure a time- and cost-efficient development in the future.

In Sec. 2 the image processing approach is described. The proper illumination concept for the recording of the knitted fabric images and the extraction of three features to classify different fabrics will be described. Furthermore, a method for the measurement of the expansion of fabric loops on image basis is shown. Additionally, the evaluation of the classification for four different knitted fabrics is shown in Sec. 3. This paper concludes in Sec. 4 and provides an outlook.

2   Approach

By a defined requirement a CPS should be able to adapt to new tasks on a low complexity level. In this case the developed system must be able to handle the switch from one fabric to another without time loss or date processing effort. To achieve this, the system needs to know the possible fabric characteristics. The first two steps are to extract features for the classification of the different fabrics and to learn a classifier to detect the right product. The third step is to estimate the expansion, as described above, for the fabric. After these steps are done, the system works like described in Fig. 2. The fabric is analysed by image processing methods and the characteristics are classified. The result of a target performance comparison can be used either for a recommendation to the operator, or for a direct command to the electric motor to speed up or to slow down.

Fig. 2. System workflow. The fabric is analysed by image processing methods. A classification is done to detect the occurring product. Then a target performance comparison is done. Finally the output results in a recommendation or an action.

2.1   Image Processing

Image processing techniques are used to retrieve features for the classification of the knitted fabric. A central aspect of every image processing task is the illumination concept. For the analysis of textiles a diffuse backlight is used [10]. This avoids shadows and reflections [11]. Furthermore, the surrounding illumination is negligible because the backlight illumination shines directly into the camera lens. The acquired image consists of dark pixels for the yarn material and bright pixels for the pores, see Fig. 3.

Firstly an image gets preprocessed by low-level image processing methods. A median filter is applied for de-noising of the image and an auto-contrast function is applied to optimise the image values [12, 13]. The segmentation of the image is carried out by a thresholding technique. The optimal threshold value is calculated by Otsu's method [14]. Due to the illumination concept, the image has a bi-modal grayscale distribution. Thus Otsu's method is suitable for this application. The result of the first steps can be seen in Fig. 3.

In the following all connected regions in the binary image need to be labelled [12, 13]. The global image information and the knowledge about the labelled regions are used to calculate features for the classification of the fabric. Theoretical foundations regarding feature extraction can be found in [15]. A pilot approach was based on the local recognition of the pore shapes. First investigations show that the pores of different fabrics are too similar for proper classification. Instead of this, the entire image is analysed. Trials showed that the number of pores, the yarn fineness, and the average pore size in the image are appropriate features to classify a knitted fabric.

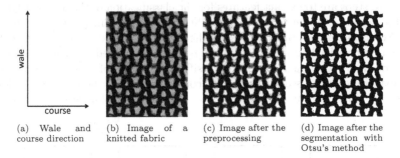

(a) Wale and course direction

(b) Image of a knitted fabric

(c) Image after the preprocessing

(d) Image after the segmentation with Otsu's method

Fig. 3. Visualisation of the first image processing steps. On the left a legend for the wale- and course-direction of the images is shown (a). The images (b)-(d) show a fabric (b), the preprocessed image of the fabric (c), and the binarised image after the segmentation (d).

Feature Extraction As described in the previous section, features for the classification task are extracted by analysing the global image. The features that are used in this work will be described in the following.

The Euler number (N_E) is the total number of connected regions in an image (N_R) excluded the number of holes (pores) in each region (N_H) [13]. It is described by the equation: $N_E = N_R - N_H$. In this application, the Euler number provides the number of pores per image. If the pore sizes of two fabrics are different, investigations show that the number of pores are different, too, see Fig. 4 (b) and (c). In this case the Euler number for both fabrics differ.

Furthermore, two different fabrics may have a similar number of pores in the image, see Fig. 4 (c) and (d), and Fig. 5 (a). In this case the yarn fineness (dark pixel) differs. The total number of black pixels (N_{BP}) in the segmented image is an indicator for the amount of yarn. It is calculated by counting the zero-elements in the m by n binary-image matrix:

$$N_{BP} = \sum_{i=1}^{m} \sum_{j=1}^{n} \{x_{ij} = 0\}. \tag{1}$$

As a third feature, the average pore size in the image is used. It can be calculated by counting the ones in the m by n binary-image matrix divided by the total number of regions (pores).

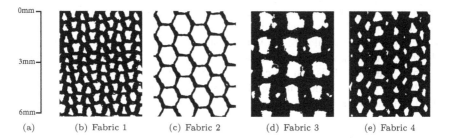

(a)   (b) Fabric 1   (c) Fabric 2   (d) Fabric 3   (e) Fabric 4

Fig. 4. Images of four different fabrics (b)-(c) with the measuring scale (a).

$$N_{AvgPS} = \frac{\sum_{i=1}^{m} \sum_{j=1}^{n} \{x_{ij} = 1\}}{N_R}.$$ (2)

These three features are used to classify the four different knitted fabrics shown in Fig. 4. Four classifiers (Naïve Bayes, support-vector machine, decision tree, 3-nearest-neighbours [16]) were tested to obtain the classification rates for the features and to estimate the best of the four classifiers. The results are shown in Sec. 3.

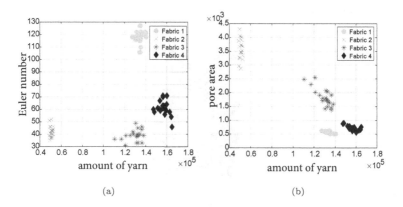

(a)   (b)

Fig. 5. Representation of classes in a 2D feature space. The left figure (a) shows the amount of yarn vs. Euler number. The right figure (b) shows the amount of yarn vs. pore area. In both cases there are four clusters, each consisting of elements of the same class that corresponds to one of the four kinds of fabric.

2.2 Expansion Measure

Once the classification of different knitted fabrics is finished, it is possible to define quality parameters for each fabric in wale direction. One potential parameter is the expansion of the fabric during production, controlled by an electric motor. If a textile material is stretched, the pore size increases in stretch direction (wale direction). Perpendicular to that direction—in course direction—the pore size decreases. The expansion of the material can be measured by the width-to-height ratio of the pore. This can be calculated by the Bounding-Box function. After the labelling of the regions in the binary image, a pore is represented as a region R. Then the Bounding Box for a pore is defined as follows [13]:

$$BoundingBox(R) = (m_{min}, m_{max}, n_{min}, n_{max}) \qquad (3)$$

where m_{min}, m_{max} and n_{min}, n_{max} are the minimal and maximal coordinates for all pixel $(x_m, x_n) \in R$. With these values it is possible to calculate the height (H) and width (W) of a pore. By dividing $\frac{H}{W}$, a value for the expansion (E) is calculated. To gain robustness, the arithmetic mean of all pores in one image is calculated instead of using only one pore, see Eq. (4). A visualisation of the method can be seen in Fig. 6 (b).

$$E = \frac{\overline{H}_{arithm}}{\overline{W}_{arithm}} \qquad (4)$$

The expansion measure is implemented on resource-efficient hardware with an Android OS. It is integrated in a demonstrator, see Fig. 6 (a).

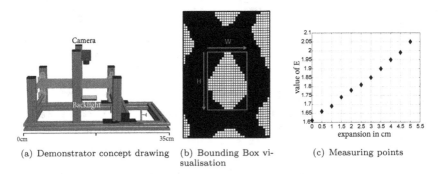

(a) Demonstrator concept drawing   (b) Bounding Box visualisation   (c) Measuring points

Fig. 6. Concept drawing of the demonstrator, side view (a). Visualisation of the expansion measure approach with a Bounding Box (b). Measurement points visualising a continuous stretched fabric, recorded with the demonstrator (c).

3 Classifier Evaluation

A test set has been created in order to evaluate the classification rates of the extracted features and to estimate the best out of four classifiers. This set contains 20 images for each of the four textiles shown in Fig. 4. According to this, there are 80 objects in total, distributed over four classes. The representation of the classes in the feature space with the calculated features is shown in Fig. 5.

For the classification of the knitted fabrics, four classifiers were tested. The learning and the classification were carried out automatically in the frame of *K-fold cv paired t test* for generalisation [16] with $K = 10$, i.e., in each fold 90% of the objects are considered for the training and the rest for the validation. The tested classifiers are Naïve Bayes, support-vector machine, decision tree and the 3-nearest-neighbours classifier [16]. The results of the evaluation can be seen in Tab. 1. The probabilistic-based Naïve Bayes and the rule-based decision tree yield the best performance.

Table 1. Classification results in the frame of *K-fold cv paired t test* for the four textiles shown in Fig. 4.

Classifier	Accuracy	Standard Deviation
Naïve Bayes	100%	+/- 0%
Support-Vector Machine	81,25%	+/- 6,25%
Decision Tree	98,75%	+/- 3,75%
3-Nearest-Neighbours	92,50%	+/- 8,29%

4 Conclusion and Outlook

This work shows a system that combines machine learning and image processing and applies it to a knitting process. It shows a suitable illumination concept for the analysis of knitted fabrics and suggests a proper method for the segmentation of images of different knitted fabrics. Furthermore, this work suggests three possible features for the classification of knitted fabrics. The extracted features are evaluated with four classifiers. In addition an approach is shown, measuring the expansion of the fabric during the production process.

As a next step it is planned to implement the classification approach shown in this paper on resource-efficient hardware. In future work the image-processing application can be extended with more functionality, e. g. defect detection. In addition to this, the concepts shown in [7–9] need to be adapted to resource efficient hardware. These two methods will be combined and coupled with a motor. It results in a seeing and feeling drive in terms of a CPS. The demonstrator shown in Fig. 6 will be furnished with an additional force meter in order to quantify the influence of fabric tension on the alternation of loop length. The implementation of the system into the inner cylinder of a large circular knitting machine at ITA is planned. In particular, the online measurement technology can be established due to the mesh structure resembling between weft and warp knitted fabrics in the field of knitted fabrics.

Acknowledgement This work is in part founded by the leading-edge cluster-project itsowl-TT-kapela, grant. no. 02PQ3062.

References

1. R. Hildebrand, J. L. Hoffmann, E. Gillich, H. Dörksen, and V. Lohweg, *Smartphones as Smart Cameras - Is It Possible?* Lemgo, Germany: inIT - Institute Industrial IT, Ostwestfalen-Lippe University of Applied Sciences, 2012.
2. V. Lohweg, J. L. Hoffmann, H. Dörksen, R. Hildebrand, E. Gillich, J. Hofmann, and J. Schaede, *Banknote Authentication with Mobile Devices*. Lemgo, Germany and Lausanne, Switzerland: inIT - Institute Industrial IT, Ostwestfalen-Lippe University of Applied Sciences, 2013.
3. K.-T. Cheng and Y.-C. Wang, *Using Mobile GPU for General-Purpose Computing - A Case Study of Face Recognition on Smartphones*. University of California, Santa Barbara, CA, USA, 2011.
4. C.-Y. Fang, W.-H. Hsu, C.-W. Ma, and S.-W. Chen, *A Vision-based Safety Driver Assistance System for Motorcycles on a Smartphone*. 2014.
5. DIN, *Textilen. Grundbegriffe*. Berlin: Beuth-Vertrieb GmbH, 1969.
6. K. P. Weber, *Wirkerei und Strickerei*. Frankfurt am Main: Deutscher Fachverlag GmbH, 2004.
7. M. Bator, A. Dicks, U. Mönks, and V. Lohweg, *Feature Extraction and Reduction Applied to Sensorless Drive Diagnosis*. Lemgo, Germany: inIT - Institute Industrial IT, Ostwestfalen-Lippe University of Applied Sciences, 2012.
8. H. Dörksen, U. Mönks, and V. Lohweg, "Fast classification in industrial Big Data environments," in *Emerging Technology and Factory Automation (ETFA), 2014 IEEE*, pp. 1–7, 2014.
9. H. Dörksen and V. Lohweg, "Automated Fuzzy Classification with Combinatorial Refinement," in *Emerging Technology and Factory Automation (ETFA), 2015 IEEE*, 2015.
10. B. Neumann, *Bildverarbeitung für Einsteiger: Programmbeispiele mit Mathcad*. Berlin Heidelberg: Springer, 1 ed., 2005.
11. A. Kumar, *Computer Vision-based Fabric Defect Detection: A Survey*. New Delhi, India: IEEE Xplore, 2008.
12. K. D. Tönnies, *Grundlagen der Bildverarbeitung*. München: Pearson Studium, 1 ed., 2005.
13. W. Burger and M. J. Burge, *Digitale Bildverarbeitung*. Berlin Heidelberg: Springer, 1 ed., 2005.
14. N. Otsu, *A Threshold Selection Method from Gray-Level Histogramms*. Tokyo, Japan: IEEE Xplore, 1979.
15. I. Guyon, S. Gunn, M. Nikravesh, and L. A. Zadeh, *Feature Extraction: Foundations and Applications (Studies in Fuzziness and Soft Computing)*. Secaucus, NJ, USA: Springer-Verlag New York, Inc, 2006.
16. E. Alpaydin, *Introduction to Machine Learning*. Cambridge, Massachusetts - London, England: The MIT Press, 2 ed., 2010.

Efficient engineering in special purpose machinery through automated control code synthesis based on a functional categorisation

Tobias Helbig[1], Steffen Henning[2], and Johannes Hoos[3]

[1] GSaME, University of Stuttgart
[2] Institute industrial IT, Lemgo
[3] Festo AG & Co.KG, Esslingen

Abstract. Individual customer demands in special purpose machinery shift the focus towards efficient engineering. However, automated engineering approaches fail due to lack of an appropriate description language for the system modelling. Therefore, a categorisation of capabilities and functions is proposed, that can serve as basis for the description of machine parts in model-based approaches. As an example an approach that synthesises control code from a plant model, based on the description language developed in this paper, is presented.

Keywords: categorisation, process function, control code synthesis, special purpose machinery

1   Introduction

Special purpose machine manufacturing is one of the key industrial sectors in Germany. As the name already suggests, and because of increasing complexity and individuality of products, each and every special purpose machine has to be developed specifically for each customer [9]. Thus the costs of engineering such systems increases and efficiency becomes the major factor of success for manufacturers [12].

Engineering includes the mechanical, electrical and software domain, which jointly develop manufacturing plants. However, each involved domain possesses its own description languages, engineering tools and data formats [1]. Thus immensely hindering seamless interaction between these domains. As a result, data that was already generated by one domain has to be (manually) recreated by the other domains. Due to a lack of a cross-domain data model, many potentials to increase efficiency of the engineering remain unused [10, 6].

This paper proposes a standardised description language to categorise machine components based on their capabilities and functions. Based on this description language a control code synthesis approach is presented. This approach uses the presented description language to automatically synthesise parts of the control code for a manufacturing plant. Thus, the approach is able to directly increase the efficiency of the engineering of special purpose machinery.

2   Analysis

The different persons involved in the engineering process have different views on the machine to be created [9]. To bridge these different views of the domains, it is suitable to use the capabilities and functions of a system as a cross-domain description language. They abstract the domain specific views on an object and serve as a common basis [6]. In the following it is analysed how capabilities and functions can be used during the engineering process.

The product to be manufactured and its process specification, defined by the costumer, are the starting base for the development of a plant. In the process view the *function* is the primary focus. It is the solution-neutral description of the relation between input, output and state variables of a system (cf. guideline VDI 2221). For example the process description defines the function "drill a hole". Here it is irrelevant how the workpiece is transported, loaded and machined, as long as the resulting hole corresponds to the specifications.

In contrast, automation components are defined by their *capabilities* and serve as the building blocks of the automation solution. The capability view describes the skills a component is able to execute, without considering its context or its effect on the product. Regarding a linear axis for example, the description of the capabilities focusses on the movement options (e.g. retract and extend). The purpose or the effect of the movement (e.g. handling or pressing) are not of interest in the capability-oriented view.

The task of the special purpose machine manufacturer is to create the process *functions*, using the available component *capabilities* [7] (cf. Figure 1). Therefore, a well-defined type of description for both views, i.e. process functions and component capabilities, is essential. An abstracted, functional model supports the creativity in the development and the understandability of the system. Furthermore it allows the model to be machine-readable, which can be used for automated engineering approaches [6]. It also facilitates re-usability of the information due to the abstract, domain-independent representation.

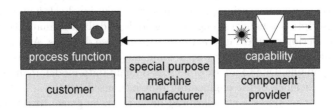

Fig. 1. Process function and capability view of the plant engineering

3   Requirements and state of the art

For a well-defined description language for process functions and component capabilities, the elements need to be categorised [6]. In conformity with [5] the following requirements apply for a categorisation:

– *Completeness*: All real world objects can be represented in the model
– *Consistency / non-redundancy*: Each real world object can be assigned to exactly one distinct representation in the model.
– *Simplicity / understandability*: The understandability of the modelled information is eased.
– *Mightiness / adequacy*: All relevant features of an object can be modelled.

A categorisation of process functions is specified in standard DIN 8580 and guideline VDI 2860. However, there are no consistent descriptions of the capabilities of automation components. The existing approaches in literature are limited to meta-models for describing capabilities in general without actually defining the distinct categories [4, 11].

In order to interrelate both views, scientific methods exist which describe the capabilities of devices of a specific sector aiming at a direct matching with process functions [3, 2]. However, these approaches assume a direct matchability between function and capability, which can rarely be found in special purpose machinery. Currently there is no seamless categorisation that meets the defined requirements and can be used as basis for automated engineering approaches.

4   Categorisation of capabilities and process functions

In this chapter the categorisation of capabilities of automation components and process functions is presented. Initially the basic ideas and the hierarchical structure of the categorisation are explained. Next, it is shown how the approach contributes to meeting the defined requirements for a categorisation. Finally the concrete categorisation for capabilities and functions is presented.

4.1   Basic concept for the categorisation

Taxonomic hierarchisation is the basic concept for the categorisation (cf. Figure 2). This means that a tree structure is applied where the categories in the top level are generic and each lower level increases the specificity. The relation between a subcategory and its parent category is defined by an "is_a-relation". This means that the subcategory maintains the features of the parent category and adds additional features. For example the category "axis" is subdivided into the subcategories "linear axis" and "rotatory axis". The subcategories maintain all features of the category "axis" and enrich it with the specification of the movement options.

Categorising an element starts with the assignment of an object to a top level category. If the object meets the specification of one of the subcategories it is further specified with the more detailed features of the corresponding subcategory. Following this procedure the categorised object can reach a very precise description with the more detailed categories in the lower levels.

Based on the taxonomic hierarchisation the categorisation is *complete*, because it can represent each possible element in a generic category at the top level. It is also *mighty*, because the detailed specification in the lower hierarchy levels can accommodate all relevant features. To guarantee the *consistency* of the model each element must be assignable to exactly one category. Therefore the subcategories are disjunct, meaning that the specification of one subcategory excludes all others. Finally, a symbolic representation of each category was designed to ease the *understandability* of the modelled information.

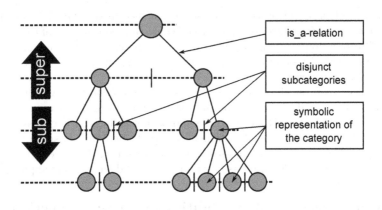

Fig. 2. Basic concept of taxonomical hierarchisation of categories

4.2   Categorisation of process functions

The process functions can be divided into three main categories: manufacture, handle and test. The category *manufacture*, specified in the standard DIN 8580, spans all functions to produce geometrically defined, solid workpieces. In the category *handling*, functions that position or orientate the workpiece are contained. The functions of the category *test* serve to verify whether workpieces fulfil distinct conditions. They are specified in detail in the guideline VDI 2860. Figure 3 shows an extract of the categorisation of process functions, representing the most important subcategories.

4.3   Categorisation of capabilities

Figure 4 represents an extract of the categorisation of the capabilities of automation components. On the top level *kinematic elements*, *handling elements*,

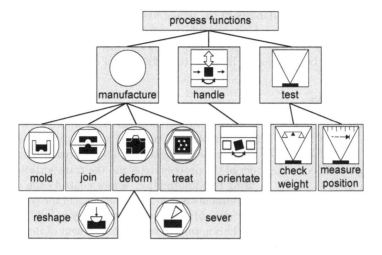

Fig. 3. Categorisation of process functions

tools, *treatment elements* and *sensors* can be distinguished. Kinematic elements can block, enable or actively move the six degrees of freedom. Handling elements serve to define and to influence the position of the material. Tools serve to induce a change of shape to the workpiece and treatment elements change the material's properties, such as its temperature. Finally sensors check whether predefined conditions are fulfilled.

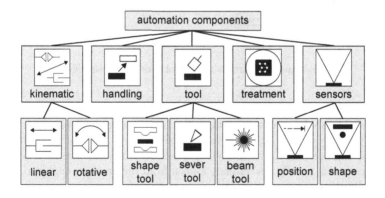

Fig. 4. Categorisation of capabilities

The presented categorisation of process functions and component capabilities was evaluated by modelling several automation solutions. Thus, proving the significance and understandability of the model based on the categorized elements. It is a well-defined and machine readable description language for the capability-oriented and the function-oriented view. Based on this, automated engineering approaches can be installed to improve the efficiency of engineering.

5   Synthesis

Especially in the software domain, automated engineering approaches are reasonable. This is because the layout plans of the mechanical engineering as well as the circuit diagrams of the electrical engineering already exist. These plans and diagrams specify the order of the process steps and the transportation paths between these steps. By utilising this information, a new control code synthesis approach is presented here. It automatically synthesises parts of the control code of a manufacturing plant (cf. Figure 5), based on a capability-based description of the plant (i.e. plant model). This approach is an enhancement of the approach described in [8].

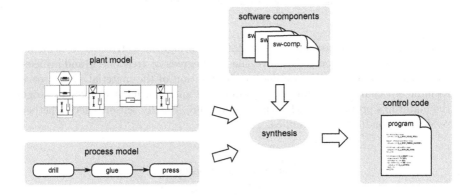

Fig. 5. Input and output of control code synthesis algorithm

The approach uses a structural plant model, whose component description is based on the capability categorisation shown in Section 4. Ideally such a model was already created during previous engineering steps, such as mechanical and electrical engineering. Additionally a process description is required (e.g. drill, glue, press). It is usually also already known from the previous engineering steps. Optionally the process can be further parameterised.

The algorithm of the control code synthesis connects these two models (plant model and process model) by assigning process steps to structural components. For example the drilling machine is assigned to the drilling process step. Then, the control code for all transport paths between the process steps is synthesised. This is done by using patterns, which represent specific plant layouts and require particular control code. The synthesis step uses the basic capabilities of the involved components. These capabilities exist as software-blocks. The result of the synthesis is executable control code for the manufacturing plant. Any desired imperative language can be specified as output format for the synthesis algorithm, e.g. IEC 61131-3 structured text (ST). Thus, the engineer can see the result in a from that he can understand and even modify, if desired.

Figure 6 shows an extract of a plant model in which the structural components for the drilling and gluing process, as well as the transport path in-between can be seen. Both components are located next to conveyor belts. These two conveyor belts are for their part connected by a handling system. After specifying the process (i.e. drill, glue), the control code for this part of the plant can be synthesised.

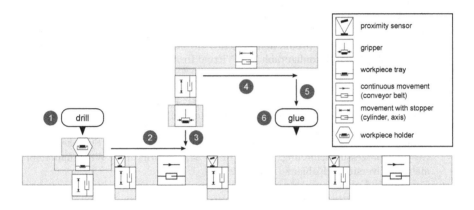

Fig. 6. Example of a plant model with process steps drill and glue. The derived process is: (1) drill; (2) move left conveyor to second stopper module; (3) grab workpiece holder; (4) move handling to right position; (5) release workpiece holder and (6) glue.

This will most notably synthesise all required control code for the transport process between these two process steps. The sequence is then synthesised as, for example ST-code. The application engineer can now view the result of the control code synthesis and decide whether or not to modify it. This drastically reduces the required programming effort and time.

Currently the algorithm is implemented in Java and can synthesise control code output in pseudo-code. However, other output formats, like ST can easily be added because of the modular structure of the synthesis program.

6 Conclusion

The lack of cross-domain models causes many redundant engineering efforts. This is because already modelled knowledge of one domain (e.g. from mechanical engineering) is currently only very rarely - if at all - used in other domains, such as software engineering. The categorisation with its symbolic representation shown in this paper offers a cross-domain means of modelling system characteristics. As an exemplary use-case this paper also presents an approach to use a plant model as input to automatically synthesise control code for that plant. This drastically reduces the required engineering time through a seamless information transfer. The algorithm then synthesises control code in a language that the user can comprehend, i.e. IEC 61131-3 structured text. This way he can easily understand and even modify the synthesised code.

References

1. Aßmann, S.: Methoden und Hilfsmittel zur abteilungsübergreifenden Projektierung komplexer Maschinen und Anlagen. dissertation, RWTH Aachen (1996)
2. Backhaus, J., Ulrich, M., Reinhart, G.: Classification, modelling and mapping of skills in automated production systems. In: 5th International Conference on Changeable, Agile, Reconfigurable and Virtual Production (CARV). Munich, Germany (2014)
3. Benkamoun, N., ElMaraghy, W., Huyet, A.L., Kouiss, K.: Architecture framework for manufacturing system design. Procedia CIRP 17 (2014)
4. Epple, U.: Austausch von Anlagenplanungsdaten auf der Grundlage von Metamodellen. Automatisierungstechnische Praxis 45 (2003)
5. Frank, U., van Laak, B.L.: Anforderungen an Sprachen zur Geschäftsprozessmodellierung. Tech. Rep. 34, Institut für Wirtschaftsinformatik, University of Koblenz-Landau (2003)
6. Gehrke, M.: Entwurf mechatronischer Systeme auf Basis von Funktionshierarchien und Systemstrukturen. dissertation, University of Paderborn (2005)
7. Henning, S., Brandenbourger, B., Helbig, T., Niggemann, O.: Plug-and-Produce für Cyber-Physische-Produktionssysteme - Eine Fallstudie im OPAK-Projekt. In: Automation 2015. Baden-Baden (2015)
8. Henning, S., Otto, J., Niggemann, O., Schriegel, S.: A descriptive engineering approach for cyber-physical systems. In: 19th IEEE International Conference on Emerging Technologies and Factory Automation (ETFA). Barcelona, Spain (2014)
9. Lindemann, U.: Methodische Entwicklung technischer Produkte. Springer-Verlag Berlin Heidelberg (2009)
10. Niggemann, O., Henning, S., Schriegel, S., Otto, J., Anis, A.: Models for adaptable automation software - an overview of plug-and-produce in industrial automation. In: Modellbasierte Entwicklung eingebetteter Systeme (MBEES). Dagstuhl, Germany (2015)
11. Rodenacker, W.G.: Methodisches Konstruieren: Grundlagen, Methodik, praktische Beispiele. Springer-Verlag Berlin Heidelberg (1991)
12. Westkämper, E.: Engineering Apps - Eine Plattform für das Engineering in der Produktionstechnik. wt Werkstattstechnik online 102(10) (2012)

Kognitive Architektur zum Konzeptlernen in technischen Systemen

Alexander Diedrich[1], Andreas Bunte[2], Alexander Maier[1], and
Oliver Niggemann[1,2]

[1] Fraunhofer Anwendungszentrum IOSB-INA, Lemgo
[2] Institut für industrielle Informationstechnik (inIT),
Hochschule Ostwestfalen-Lippe, Lemgo

Abstract. Durch die Komplexität technischer Systeme und die benötig-
te Flexibilität für Systemänderungen werden innovative Ansätze benötigt
die schnell an neue Situationen angepasst werden können. In dieser Arbeit wird die Implementierung der Architektur CATS (Cognitive Architecture for Technical Systems) beschrieben. Im Vergleich mit bestehen
Architekturen existieren einige Unterschiede: Die Kommunikation zwischen System und Bediener wird über natürliche Sprache realisiert, wobei
die Sprachverarbeitung auf dem automatischen Lernen von Konzepten
besteht. Da technische Systeme ihr Verhalten über die Zeit ändern, kann
CATS nicht auf statische Textdokumente als Wissensbasis zurückgreifen.
Stattdessen müssen die nötigen Informationen aus Echtzeitdaten extrahiert werden.
Um die Funktion von CATS zu validieren, wurde die Architektur auf
einem Demonstrator implementiert. Das Stanford CoreNLP Framework
wurde darin zur Sprachverarbeitung benutzt. Die Wissensmodellierung
wird durch die Web Ontology Language (OWL) realisiert und die Vorver-
arbeitung der Maschinendaten mit verschiedenen maschinellen Lernalgorithmen implementiert. Der Demonstrator ist in der Lage 84% der
Eingabetexte richtig zu beantworten.
Durch die Benutzung von CATS in technischen Systemen kann der In-
stallations- und Adaptionsaufwand reduziert werden, da Ansätze für
maschinelles Lernen Informationen über das System automatisch gewin-
nen. Weiterhin wird der Informationsaustausch zwischen Bediener und
Maschine durch natürliche Sprache und Konzepte vereinfacht.

1 Einleitung

Technische Systeme werden immer flexibler und modularer, wodurch die Komplexität dieser Systeme steigt [1]. Weiterhin wird die Komplexität durch adaptive
Technologien wie Industrial Internet [2] und Industrie 4.0 [3] erhöht. Benutzer
sind mit der Überwachung dieser zunehmend komplexen Systeme jedoch immer
mehr überfordert. Daher werden für die Überwachung häufig proprietäre Diagnosesysteme eingesetzt um Benutzer zu unterstützen. Die Wandelbarkeit der

technischen Systeme bedingt jedoch einen hohen manuellen Aufwand zum Anpassen der Diagnosesysteme. Maschinelle Lernverfahren können eingesetzt werden, um den manuellen Anpassungsaufwand zu minimieren und können beim Identifizieren von Fehlern sowie deren Ursachen unterstützen.

Maschinelle Lernverfahren arbeiten auf subsymbolischer Ebene (d.h. Signale über Zeit) und identifizieren Unregelmäßigkeiten oder Korrelationen in Echtzeitdaten. Menschen denken symbolisch d.h. in Konzepten [4] wie z.B. *Fehlerursache* oder *Störungsmeldung* und sind damit Maschinen in der subsymbolischen Verarbeitung unterlegen. Um Menschen effektiv unterstützen zu können, muss diese Lücke zwischen symbolischer und subsymbolischer Verarbeitung geschlossen werden. Hierfür wird eine Architektur benötigt, um in einem standardisierten Verfahren die Konzepte aus subsymbolischen Daten abzuleiten und so dem Menschen reproduzierbar Diagnosemeldungen auf symbolischer Ebene zur Verfügung zu stellen.

Ein Konzept ist eine mentale Repräsentation von Objekten oder Fähigkeiten. Konzepte werden von Menschen genutzt um die sich verändernde Welt in diskreten Kategorien darzustellen [5]. Dieses ermöglicht eine effiziente Kommunikation, da ein gemeinsames Verständnis über die Begriffe besteht und diese nicht näher erläutert werden müssen. Abhängig vom individuellen Hintergrund eines Menschen sind hierbei verschiedene Abstraktionsstufen nötig. Ein Fabrikleiter ist zum Beispiel an dem Durchsatz einer Anlage interessiert, während ein Wartungsingenieur eher Maschinenausfälle analysiert.

Um auf komplexe und benutzerabhängige Schnittstellen zu verzichten, bietet sich eine allgemeine Sprachverarbeitung an. Dabei kann der Benutzer eine Frage oder einen Befehl eingeben und bekommt eine entsprechende Antwort. Das hat den Vorteil, dass der Benutzer nicht durch Bedienoberflächen suchen muss, um die Antwort zu finden, sondern in seiner gewohnten Sprache mit der Maschine kommunizieren kann.

Solche Systeme sind bereits aus dem Konsumerbreich bekannt, wie z.B: Siri von Apple, Google Voice Search oder IBM Watson. Der Unterschied zwischen dem Konsumerbereich und einem technischen Diagnosesystem ist, dass es im Bereich der Diagnose keine Wissensbasis aus Textdokumenten gibt, aus denen die Antworten anhand von Zusammenhängen nach Wahrscheinlichkeit ermittelt werden. Deshalb sind neue Techniken erforderlich, um Antworten aus Maschinendaten zu generieren.

Dazu wurde eine kognitive Architektur (CATS) entwickelt, basierend auf der Referenzarchitektur von [6], um den oben genannten Ansprüchen gerecht zu werden. Zum ersten Mal werden gelernte Konzepte dazu benutzt, den Zustand eines technischen Systems dem Benutzer in natürlicher Sprache zu kommunizieren. Diese Arbeit zeigt, wie CATS eingesetzt werden kann, um ein technisches System zu diagnostizieren. Als Anwendungsbeispiel dient die Lemgoer Modellfabrik.

2 Kognitive Architektur zur Diagnose technischer Systeme

Dieses Kapitel beschreibt die grundlegende Funktionsweise von CATS. Im Unterschied zu den meisten anderen kognitiven Architekturen ist CATS nicht zielorientiert, sondern reagiert auf Benutzeranfragen. In Abbildung 1 ist die Architektur dargestellt. Sie besteht aus folgenden drei Hauptteilen:

- **Sprachverarbeitung (NLP - Natural Language Processing)** ist die Schnittstelle zum Menschen. Sie extrahiert die Informationen aus einem Eingabetext und stellt diesen formal, in Form eines semantischen Frames, dar. Dieser Frame wird an die Zentrale Ausführungseinheit übergeben.
- **Zentrale Ausführungseinheit (ZA)** mit dem angeschlossenen Langzeitspeicher hat zwei wesentliche Aufgaben. Zum einen werden die maschinellen Lernalgorithmen überwacht und geprüft, in welchem Zustand sich das System befindet und ob mögliche neue Konzepte ermittelt wurden. Zum anderen werden eingehende semantische Frames bearbeitet, d.h. es wird eine Anfrage an die Langzeitspeicher gestellt und das Ergebnis der Sprachverarbeitung zurückgegeben.
- **Maschinelles Lernen (ML)** beinhaltet zahlreiche Algorithmen welche die Prozessdaten nach Unregelmäßigkeiten, Anomalien und Zuständen durchsuchen.

Im Vergleich zu der Referenzarchitektur in [6] entspricht die NLP Schicht der anwendungsspezifischen Benutzerschnittstelle. Die ZA und der Langzeitspeicher bilden die konzeptuelle Schicht und repräsentieren und lernen neue Konzepte. Die ML Schicht umfasst die Lern- und Adaptierungsschichten der Referenzarchitektur und stellt ein einheitliches Interface zu den Algorithmen zur Verfügung.

2.1 Sprachverarbeitung

Benutzer interagieren mit der Sprachverarbeitungsschicht über eine Benutzerschnittstelle. Das Ziel ist, die Eingabe zu verarbeiten, sodass sie in einem semantischen Frame [7] dargestellt werden kann. Der Frame beinhaltet somit die generellen Konzepte, welche im Eingabetext vorhanden sind und stellt sie der ZA zur Verfügung. Der gesamte semantische Frame besteht aus verschiedenen Slots von denen jeder anwendungsfallspezifische Informationen enthält und durch explizite Regeln beschrieben wird. Um festzustellen, welche Daten bzw. Slots der Frame für Diagnoseaufgaben enthalten muss, wurde ein Korpus mit 150 Texten ermittelt. Daraus wurde der semantische Frame mit den folgenden sechs Slots entwickelt:

1. *Art der Eingabe*: Spezifiziert, ob der Benutzer eine Anfrage, ein Kommando oder eine Statusinformation formuliert hat.
2. *Art der Ausgabe*: Spezifiziert welche Art der Ausgabe der Benutzer erwartet (z.B. kontinuierlicher Wert, Boolscher-Wert oder Beschreibung).

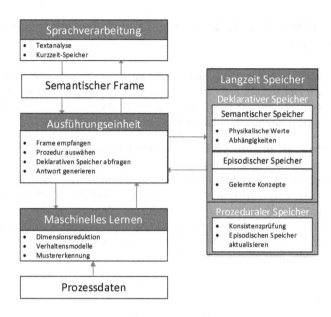

Fig. 1. Kognitive Architektur zur Diagnose von technischen Systemen.

3. *Intention*: Beschreibt den physikalischen Wert oder andere Grundkonzepte, wie ANOMALIE oder SYSTEM, welche angefragt werden.
4. *Entitätstyp*: Beschreibt den Typ der angegebenen Entität wie zum Beispiel "Motor".
5. *Entitäts ID*: Repräsentiert eine optionale alpha-numerische ID.
6. *Zeitstempel*: Gibt an, welche Zeitangabe im Eingabetext verwendet wurde [8].

Slots 1 bis 3 dienen dazu den Eingabetext in Kategorien einzuordnen. Die Kategorien werden von der ZA benutzt, um die richtigen Prozeduren aus dem prozeduralen Speicher auszuwählen. Slots 4 und 5 enthalten Inhalte die spezifisch für das jeweilige Produktionssystem sind (z.B. Gerätekategorien, Modulbezeichnungen und IDs etc.). Der Zeitstempel gibt an, auf welchen Zeitpunkt sich die Werte der anderen Slots beziehen. Die NLP Schicht hält die letzten Eingabetexte in einem Kurzzeitspeicher um Referenzen zwischen den Texten aufzulösen.

2.2   Zentrale Ausführungseinheit

Die zentrale Ausführungseinheit (ZA) ist die Hauptkomponente für die Informationsverarbeitung. Die ZA verwaltet die Kommunikation zwischen der ML- und NLP-Schicht, sowie dem Langzeit-Speicher. Die Neurowissenschaften teilen das menschliche Langzeitgedächtnis in einen deklarativen Speicher und einen prozeduralen Speicher auf, diese Aufteilung ist auch für technische Systeme sinnvoll.

Deklarativer Speicher Der deklarative Speicher umfasst Wissen über die externe Welt wie z.B. Fakten, Ereignisse, Relationen und Konzepte [9]. Menschen können dieses Wissen nutzen, um Informationen auszutauschen. Der deklarative Speicher kann in den semantischen und den episodischen Speicher aufgeteilt werden [10, 11].

Der **semantische Speicher** ist der Teil des deklarativen Speicher in dem Fakten, Relationen und Konzepte gespeichert werden. Generell enthält der deklarative Speicher Informationen die nicht auf persönlicher Erfahrung basieren [10]. Den meisten Menschen ist beispielsweise bekannt, dass London eine Stadt ist, auch wenn sie noch nicht dort waren. Daher ist dieses Faktum ein Teil des semantischen Speichers und somit auch teilbar, d.h. es kann anderen mitgeteilt werden. Das selbe gilt für Relationen wie zum Beispiel, dass London in England liegt.

In CATS enthält der semantische Speicher Informationen über die Systemstruktur, sowie über verfügbare Algorithmen und deren Anforderungen und Grenzen. Wenn eine Anfrage über zeitliche Fehler im System vorliegt, wird der semantische Speicher durchsucht und ermittelt, dass dazu z.B. ein gelernter zeitbehafteter Automat verwendet werden sollte.

Da dieses Wissen in der Domäne technischer Systeme gleich ist, wird es möglich es einmalig zu formalisieren und zwischen allen System zu teilen. Neue Algorithmen können schnell integriert werden, indem ihre Charakteristika in die Wissensbasis aufgenommen werden.

Im biologischen Umfeld enthält der **episodische Speicher** Ereignisse und Erfahrungen die bisher erlebt wurden. Dieses beinhaltet Erinnerungen an Personen, Objekte und Ereignisse zu bestimmten Zeitpunkten [10]. In CATS wird episodischer Speicher genutzt, um all die Informationen zu speichern die während der Laufzeit auftreten. Jedes neue Konzept wird in den episodischen Speicher aufgenommen. In diesem Fall ist es außerdem sinnvoll diskrete Signale abzuspeichern, damit das System nicht bei jeder Anfrage die ML Schicht abfragen muss.

Prozeduraler Speicher Im Gegensatz zum deklarativen Speicher, welcher Fakten und Ereignisse enthält, speichert der prozedurale Speicher "wie" Dinge getan werden. Menschen erhalten dieses Wissen durch wiederholtes Training. Dadurch ist dieses Wissen nicht teilbar. Beispiele hierfür sind Fahrradfahren, Schrauben anziehen und Schwimmen.

CATS verwendet diesen Speicher um anzugeben, wie das System Wissen im deklarativen Speicher verwenden kann. Da es nicht sinnvoll ist, dass das System alle Prozeduren lernt, wird dieses Wissen manuell modelliert. D.h. wenn ein semantischer Frame gegeben ist, wählt die ZA eine Prozedur aus, welche die Information im semantischen Frame sinnvoll bearbeiten kann. Prozeduren beinhalten Schritte die abzuarbeiten sind, um eine Antwort zu generieren. Die Herausforderung besteht darin, mit den Prozeduren die richtige Abfrage des semantischen Speichers zu generieren.

2.3 Maschinelles Lernen

Die ML Schicht stellt sub-symbolische Funktionen für die Identifikation von Verhaltensmodellen, Mustererkennung und Anomalieerkennung zur Verfügung. Die ML Schicht hat Zugriff auf die Prozessdaten und entkoppelt die ZA von den unterliegenden Prozessen. Aufrufe der ML Schicht werden immer von der ZA initiiert.

Es werden verschiedene maschinelle Lernverfahren zur Datenanalyse verwendet. Da CATS lediglich eine Architektur beschreibt und unabhängig von den konkreten unterliegenden Prozessen ist, wird nicht spezifiziert welche ML Algorithmen verwendet werden sollen. Beispiele für verschiedene Algorithmen zum identifizieren von sub-symbolischen Informationen sind:

1. PCA zur Dimensionsreduktion der Eingangsdaten [12].
2. Regel-basiertes Schließen [13].
3. HyBUTLA [14] zur Identifizierung von (hybriden) zeitbehafteten Automaten

Durch ihre Struktur (Zustände und Transitionen mit Ereignissen) sind zeitbehaftete Automaten besonders geeignet um Konzepte abzuleiten. Vereinfacht dargestellt wird jeder Zustand als ein Konzept betrachtet, da Zustände das Verhalten des Systems angeben. Dieses ist ähnlich der menschlichen Art und Weise Konzepte zu erkennen.

2.4 Konzeptlernen

Als zentrales Element ruft die ZA alle Algorithmen periodisch auf. Die Ausgaben der Algorithmen werden in den episodischen Speicher geschrieben. Sobald ein Algorithmus einen neuen Status erkennt, wird dieser als neues Konzept aufgefasst. Dieses Vorgehen ähnelt der menschlichen Art und Weise Konzepte zu lernen. Der Unterschied ist, dass durch die begrenzte Kapazität des menschlichen Gehirns, Menschen nicht jeden einzelnen neuen Zustand erfassen. Stattdessen speichern Menschen nur die Zustände, die sie für wichtig erachten. Durch hohe Verarbeitungsgeschwindigkeiten und verfügbaren Speicher ist es Maschinen möglich jeden einzelnen neuen Zustand abzuspeichern. Somit kann CATS ein detaillierteres Modell eines Produktionssystems erstellen als Menschen.

Um ein Konzept zu klassifizieren werden die Ausgaben von verschiedenen Lernalgorithmen benutzt. Wenn eine neue Beobachtung erkannt wurde, wird diese mit den bisher gespeicherten Konzepten durch ein Distanzmaß verglichen. Die Distanz gibt die attributweise Differenz zwischen der neuen Beobachtung und den durchschnittlichen Attributwerten jedes Konzeptes an. Um zu überprüfen, ob die Beobachtung zu einem bereits bestehenden Konzept gehört oder ob es sich um ein neues Konzept handelt, wird die Distanz zu den bestehenden Konzepten berechnet. Die Beobachtung wird einem bestehenden Konzept zugeordnet, wenn die Distanz zu diesem Konzept kleiner als ein Schwellwert ist. Ansonsten wird ein neues Konzept erstellt und die Beobachtung wird zum ersten Ereignis interpretiert.

Die Konzepte werden zusammen mit Meta-Informationen im episodischen Speicher abgelegt. Die Meta-Information können zur statistischen Analyse benutzt werden, z. B. um langsame Änderungen im Produktionssystem zu erkennen. Weiterhin können die Meta-Informationen Daten umfassen die von Menschen eingegeben werden wie z.b. Name des Konzeptes oder Anweisungen. Das Festlegen eines Namens für ein Konzept ist essentiell, da dieses erst die Grundlage zur Kommunikation zwischen Mensch und Maschine darstellt.

3   Anwendungsbeispiel

Um die Praxistauglichkeit der Architektur zu evaluieren, wurde die Architektur auf einem Demonstrator in der Lemgoer Modellfabrik (Abbildung 2) implementiert. Als Beispieleingabe wird der Satz "Ist das System ok?" verwendet.

Zur Sprachverarbeitung wird das Stanford CoreNLP Framework eingesetzt [15]. Slots 1 bis 3 des semantischen Frames werden durch Klassifikation bestimmt. Slots 4 und 5 können nicht klassifiziert werden, da die möglichen Werte vom eingesetzten Produktionssystem abhängig sind. Daher werden diese über ein zweistufiges Verfahren ermittelt: Zuerst wird der Eingabetext mit einem Stemmer und einem Lemmatizer vorverarbeitet. Danach versucht das NLP Framework die beiden Slots durch reguläre Ausdrücke zu ermitteln. Die regulären Ausdrücke akzeptieren alle Entitäten die eine entsprechende alphanumerische ID haben. Wird kein Ausdruck akzeptiert, werden die Entitäten durch part-of-speech (POS) tagging und Abhängigkeitsanalyse ermittelt. Für den hier dargestellten Fall wird Slot 4 über POS tags ermittelt, da es keine ID gibt. Slot 6 wird mit dem SUTime Framework [8] gefüllt. Da in dem Beispiel keine Zeitinformation vorhanden ist, wird der Präsens angenommen. Insgesamt ergibt sich der folgende semantische Frame:

$$SF = (Status, Boolean, System, System, NULL, present\_ref)$$

In diesem Fall wird nach einer Statusinformation gefragt, wobei eine boolesche Antwort erwartet wird (System ist ok bzw. nicht ok). Die Intention und die Entität beziehen sich auf das System als Ganzes.

Die ZA verarbeitet diesen semantischen Frame und wählt eine Prozedur aus dem prozeduralen Speicher. Jede Prozedur beinhaltet Templates (zur Abfrage des Langzeitspeichers), welche mit den Inhalten des semantischen Frames gefüllt werden. Das Ziel hierbei ist, die Konzepte des Eingabetextes zu erkennen (im Beispiel STATUS). Ein Problem ist, dass mit dieser Art nur Konzepte gefunden werden, deren Namen im semantischen Frame und im episodischen Speicher exakt gleich sind. Da dieses jedoch nicht immer der Fall ist, wird die Levensthein Distanz (auch bekannt als die Editierdistanz) verwendet, um z.B. Rechtschreibfehler zu korrigieren. Dadurch wird die Benutzerfreundlichkeit des Systems signifikant erhöht.

Der semantische Speicher wird durch eine OWL Ontologie repräsentiert. Jedes Konzept ist eine Klasse in der Ontologie, wie zum Beispiel STATUS. Abhängigkeiten zwischen den Klassen werden durch Objekteigenschaften model-

Fig. 2. Demonstrator in der Lemgoer Modellfabrik.

liert. Diese Konzepte und Relationen sind in der gesamten Domäne von technischen Systemen gleich.

Die Inhalte des episodischen Speichers werden durch Instanzen von OWL Klassen repräsentiert. Individuals repräsentieren konkrete Dinge in der realen Welt. Daher sind alle Individuals Instanzen von OWL Klassen und repräsentieren ein konkretes Konzept. Um eine Antwort zu finden, wird das Konzept der Antwort identifiziert und alle Individuals dieses Konzeptes zurückgegeben. Die Individuals werden nach ihrem Status sortiert und historisch geordnet.

In der ML Schicht wird der OTALA Algorithmus [16] verwendet, um Verhaltensmodelle als zeitbehaftete Automaten zu identifizieren. Da OTALA nur diskrete Zustände des Systemverhaltens abbildet, wurde zusätzlich eine PCA [17] implementiert, um die Verarbeitung von kontinuierlichen Signalen zu ermöglichen. Der Demonstrator hat zwei kontinuierliche Signale, Leistung und Energie.

Der automatisch identifizierte Automat (abgeleitet aus dem Demonstrator) umfasst sechs Zustände. In jedem Zustand sind die aktiven Signale bekannt. Somit können Konzepte wie "Förderband läuft nach rechts" oder "Objekt hat Position A erreicht" gelernt werden. Um Informationen über die kontinuierlichen Signale (z.B. den normalen Energieverbrauch) zu erhalten wird die PCA benutzt. Während der Laufzeit wird diese Information benutzt, um die Wahrscheinlichkeit zu berechnen, dass der derzeitige Energieverbrauch normal ist. Somit können Konzepte wie "korrekter Energieverbrauch in Zustand S2" gelernt werden.

Um Missverständnisse zu vermeiden, enthält die generierte Antwort detaillierte Informationen über die ausgeführten Schritte. Dadurch wird der Benutzer in die Lage versetzt zu überprüfen, ob der Eingabetext richtig verstanden wurde. Basierend auf der Antwort erstellt die NLP Schicht einen Antworttext in natürlicher Sprache. Wenn keine Anomalien aufgetreten sind, ist die Antwort "Es gibt keine Anomalien im System". Wenn die PCA eine geringe Wahrscheinlichkeit für einen korrekten Energieverbrauch in Zustand 4 berechnet hat, ist die Antwort "Der Energieverbrauch in Zustand S4 ist anormal". Wenn zusätzlich bekannt ist, dass Zustand 4 die Bewegung des Laufbandes beschreibt, wäre die Antwort "Es tritt ein anormaler Energieverbrauch während der Operation des Laufbandes auf".

Validierung Um den CATS Prototypen zu validieren, wurden 150 Beispiel-texte von Maschinenbedienern formuliert. Tabelle 1 zeigt die Ergebnisse der Validierung. Die Texte wurden cross-validiert mit 10 % Test- und 90 % Trainings-daten. Positive Antworten sind diagonal dargestellt mit 104 true-positive (TP) und 22 true-negative (TN) Werten. Die TP Werte geben an, dass eine korrekte Antwort generiert wurde bei einem unterstützten Eingabetext. TN Werte geben an, dass der Eingabetext nicht unterstützt wurde, das System dieses jedoch erkannt hat und eine verständliche Antwort generiert hat. False-positive (FP) Werte sind in der links-unteren Zelle in Tabelle 1 eingetragen. Diese geben an, dass auf einen unterstützten Eingabetext eine falsche Antwort erfolgt ist. Drei FP Werte kamen durch einen inkorrekten semantischen Frame zustande. Die restlichen FP Werte traten durch falsche Verarbeitung in der ZA auf. Im Ganzen wurde eine Akkuranz von 84% und ein F1-Score von 89,6% erreicht.

	Referenz		Akkuranz	F1-Score
	True	False		
Beobachtung true	104	0	84,0%	89,6%
Beobachtung false	24	22		

Table 1. Ergebnisse des CATS Prototypen in der Lemgoer Modellfabrik

4 Zusammenfassung

In dieser Arbeit wurde gezeigt, wie die Diagnose von technischen Systemen durch Konzeptlernen erreicht werden kann. Die Übersetzung zwischen sub-symbolischen Daten in eine symbolische Repräsentation durch Konzepte wird in einer konzeptu-ellen Schicht mit einer zentralen Ausführungseinheit und Langzeitspeicher realisiert. Das Lernen von Konzepten geschieht, indem eine neue sub-symbolische Eingabe dem Konzept zugeordnet wird, zu dem es die größte Ähnlichkeit hat. Wenn kein Konzept ähnlich genug ist, wird ein neues Konzept erstellt. Insgesamt wurde die konzeptuelle Schicht in einer drei-Schichten Architektur verwendet, um die Informationsverarbeitung zu standardisieren.

CATS wurde in der Lemgoer Modellfabrik validiert. Weiterhin wurden implementierungsspezifische Details an einem Beispiel demonstriert. Wenn Bediener eine Implementierung von CATS benutzen, können sie Fragen und Kommandos in natürlicher Sprache eingeben, ohne ein komplexes Benutzerinterface bedienen zu müssen.

Zukünftige Forschungsarbeiten könnten darin bestehen Subsysteme zu erkennen um eine genauere Diagnose ermöglichen, die Wissensbasis zu erweitern oder CATS auch für andere Industriezweige nutzbar zu machen (z.B. in der Prozessindustrie). Zusätzlich könnten weitere maschinelle Lernverfahren die Art von diagnostizierbaren Fehlern vergrößern. Außerdem könnten mehrere semantische Frames definiert werden, um mehr Funktionalitäten zu unterstützen und einzelne semantische Frames zu spezialisieren.

5   Danksagung

Das dieser Veröffentlichung zugrundeliegende Projekt "Semantics4Automation" wird mit Mitteln des Bundesministeriums für Bildung und Forschung unter dem Förderkennzeichen 03FH020I3 gefördert.

References

1. R. Berger, "Mastering product complexity," 2012.
2. P. C. Evans and M. Annunziata, "Industrial internet, pushing the boundaries of minds and machines," 2012.
3. H. Kagermann et. al., "Recommendations for implementing the strategic initiative industrie 4.0," *Frankfurt/Main, National Academy of Science and Engineering*, 2013.
4. T. D. Kelley, "Symbolic and sub-symbolic representations in computational models of human cognition, what can be learned from biology?" in *Theory & Psychology*, vol. 13, no. 6, 2003, pp. 847–860.
5. B. M. Lake, "Towards more human-like concept learning in machines: Compositionality, causality, and learning-to-learn," Ph.D. dissertation, Massachusetts Institute of Technology, 2014.
6. O. Niggemann, G. Biswas, J. S. Kinnebrew, H. Khorasgani, S. Volgmann, and A. Bunte, "Data-driven monitoring of cyber-physical systems leveraging on big data and the internet-of-things for diagnosis and control," 2015.
7. A. Moschitti, P. Morarescu, and S. M. Harabagiu, "Open domain information extraction via automatic semantic labeling." in *FLAIRS Conference*, 2003, pp. 397–401.
8. A. X. Chang and C. D. Manning, "Sutime: A library for recognizing and normalizing time expressions." in *LREC*, 2012, pp. 3735–3740.
9. E. Tulving, W. Donaldson, and G. H. Bower, *Organization of Memory*, E. Tulving, W. Donaldson, and G. H. Bower, Eds.   Academic Press, 1972.
10. E. Yee, E. G. Chrysikou, and S. L. Thompson-Schill, *The Oxford Handbook of Cognitive Neuroscience*, K. Ochsner and S. M. Kosslyn, Eds.   Oxford University Press, 2013, vol. 1.
11. A. K. Krupa, "The competitive nature of declarative and nondeclarative memory systems: Converging evidence from animal and human brain studies," *The UCLA USJ*, vol. 22, 2009.

12. T. Chen and J. Zhang, "On-line multivariate statistical monitoring of batch processes using gaussian mixture model," *Computers and Chemical Engineering*, vol. 34, no. 4, pp. 500 – 507, 2010. [Online]. Available: http://www.sciencedirect.com/science/article/pii/S009813540900218X

13. L. F. Mendonça, J. M. C. Sousa, and J. M. G. Sá da Costa, "An architecture for fault detection and isolation based on fuzzy methods," *Expert Syst. Appl.*, vol. 36, no. 2, pp. 1092–1104, Mar. 2009. [Online]. Available: http://dx.doi.org/10.1016/j.eswa.2007.11.009

14. O. Niggemann, B. Stein, A. Vodenčarević, A. Maier, and H. Kleine Büning, "Learning behavior models for hybrid timed systems," in *Twenty-Sixth Conference on Artificial Intelligence (AAAI-12)*, Toronto, Ontario, Canada, 2012, pp. 1083–1090.

15. C. D. Manning, M. Surdeanu, J. Bauer, J. Finkel, S. J. Bethard, and D. McClosky, "The stanford corenlp natural language processing toolkit," in *Proceedings of 52nd Annual Meeting of the Association for Computational Linguistics: System Demonstrations*, 2014, pp. 55–60.

16. A. Maier, "Online passive learning of timed automata for cyber-physical production systems," in *The 12th IEEE International Conference on Industrial Informatics (INDIN 2014)*. Porto Alegre, Brazil, Jul 2014.

17. J. Eickmeyer, P. Li, O. Givehchi, F. Pethig, and O. Niggemann, "Data driven modeling for system-level condition monitoring on wind power plants," 2015, presented at the 26th International Workshop on Principles of Diagnosis (DX-2015).

Implementation and Comparison of Cluster-Based PSO Extensions in Hybrid Settings with Efficient Approximation

André Mueß[1], Jens Weber[1], Raphael-Elias Reisch[2], and Benjamin Jurke[3]

[1]Heinz Nixdorf Institut, Wirtschaftsinformatik, insb. CIM, Paderborn, Germany
amuess@mail.upb.de, jens.weber@hni.upb.de
[2]Fachhochschule Bielefeld, Fachbereich Ingenieurwissenschaften und Mathematik,
Bielefeld, Germany
raphael-elias.reisch@fh-bielefeld.de
[3]DMG MORI AG, Bielefeld, Germany
benjamin.jurke@dmgmori.com

Abstract. This contribution presents a comparison between two extensions of the particle swarm optimization algorithm in a hybrid setting where the evaluation of the objective function requires a high computational effort. A first approach using simulation-based optimization via particle swarm optimization was developed in order to reach an improved setup optimization support of the workpiece position and orientation in a CNC tooling machine. For that, a 1:1 interface between the machine simulation model and the simulation-based optimization approach produced a high number of simulation runs. The idea arose that the extension of the PSO algorithm as well as the usage of an NC interpreter operating as a pre-processing component could support the setup process of the tooling machines. The extension of the PSO algorithm deals with the segmentation of the parameter search space taking collisions and lower computational effort into consideration. A significant reduction of simulation runs has been achieved.

Keywords: NC interpreter, K-means, binary search, workpiece, tooling machine, clustering, PSO extension

1   Introduction

In order to provide optimal tooling machine parameters as well as production parameters via simulation, the idea came up to create an automatic setup optimization process using simulation-based optimization methods modeled after [4, 5]. A simulation model of the tooling machine serving to evaluate the given parameters and a swarm-based metaheuristic (Particle Swarm Optimization, abbreviated as PSO) serving as the optimization component builds the core of the approach. The goal is to reach near optimal parameters for the positions of workpiece, workpiece clamp and it offers further setup parameters in a multi-objective environment. In this parameter optimization

adfa, p. 1, 2011.

process, especially using experimental parameters, a manual search of useful parameters is necessary and it typically requires an impractical effort that can be handled by methods of design of experiments (DOE) [6].

Additionally, with the simulation-based optimization process each particle acting as a solution candidate for the setup parameters requires a single simulation run. This leads to a high number of simulations as well as a high computational costs, even if the simulations are able to run in parallel. The contribution of [7] offers initial PSO extension approaches in order to improve the particle swarm algorithm via the implementation of asynchronous and partially synchronous features to handle the limited resources and parallel optimization processes. In specific settings, the evaluation of each solution parameter set is computationally expensive and therefore limited both in terms of the size of the swarm and the number of generations. In a situation where computation time and the number of raw simulation runs are expensive to reach an adequate result, an approximation method with negligible costs is available. A high number of approximation runs may be used to analyze the search space for nonviable regions in order to reduce the number of required computational runs. In order to determine the solution candidates for the setup optimizing process based on this approximation approach, either a K-means clustering (according to [8]) or a binary search algorithm (according to [1]) is used to preselect data points and reduce the search space.

In this contribution, both approaches are presented and compared. In section two, the formalization of the PSO extensions is shown and the developed concept is described and illustrated. Section three deals with the description of the developed experimental design. Section four discusses the findings of the experiments and Section five closes the submission with a conclusion and outlook.

2   Formalization and Representation of the PSO Extension approach

Let $f: \mathbb{R}^d \to \mathbb{R} \times \{0,1\}$ be a fitness function which both denotes the fitness value and the validity of a given input. More specifically, $g: \mathbb{R}^d \to \mathbb{R}$ denotes a function that just computes a fitness value whereas $h: \mathbb{R}^d \to \{0,1\}$ is to compute the validity of a given input vector. We assume that the computation of g requires significantly less time than the computation of h. Both approaches presented in this paper consist of two phases where the first phase is to figure out a selection of parameter vectors having a satisfying fitness value. This is achieved by performing a number of PSO iterations considering (according to [2, 3]) g being the fitness function. Let $X := \{(a, b): a \in \mathbb{R}^d, b \in \mathbb{R}, b = g(a)\}$ be the collection of possible solutions which have been computed during this search procedure. The second step of the algorithm is supposed to divide the computed data X. Let k be the number of available computing resources. Furthermore, let $S := (s_1, s_2 \ldots s_k)$ where $s_i \subseteq X$ with $a \in \mathbb{R}^d$ and $b \in \mathbb{R}$ where, as before, a denotes the input vector and $b = g(a)$. We assume that each s_i is ordered by b. In this contribution, S is computed in two different ways. The first approach is to employ K-means with X functioning as the input data and k denoting the number of clusters. The second approach is an adapted binary search technique which is just described for 2-dimensional problems. It proceeds as follows: Let $l_i = \max(a_i) - \min(a_i)$ be the expansion of each direction in a. In this case, we assume that $S := (s_{1,1}, \dots, s_{m,n})$ where $m * n = k$. Additionally, $|m - n|$ is to be minimal. The assignment is achieved by

$$S_{i,j} := \left\{ (a,b) \in X \mid \frac{l_x}{m} * i \leq a < \frac{l_x}{m} * (i+1) \text{ and } \frac{l_y}{n} * j \leq b < \frac{l_y}{n} * (j+1) \right\}$$

$$= \left(\left(\frac{l_x}{m} * [i, i+1] \right) \times \left(\frac{l_y}{n} * [j, j+1] \right) \right) \cap X.$$

Algorithm 1 shows how the values m and n are computed. Based on these decompositions, the search for the best valid input vector is performed by an elimination technique. At first, the vectors with the best fitness values out of each $s_i \in S$ are considered. Afterwards, for each $i \in \{1, \dots, k\}$ $h(b_i)$ is computed concurrently. If $h(b_i)$ is valid, each candidate (a, b) out of each cluster s_j with $b > b_i$ is deleted. While any cluster contains a candidate which has not been validated or deleted, the procedure iterates through it until a valid solution is found.

Input: Number of areas k
if k is a prime number
    $k \leftarrow k - 1$;
endif
for $(i = \sqrt{k} \rightarrow 2; i = i - 1)$
    if $k \bmod i = 0$
        $m \leftarrow i$;
        $n \leftarrow k/i$;
    endif

endfor

Algorithm 1. Decomposition values for the adapted binary search procedure

2.1 System Concept

The NC interpreter is able to approximate the duration time of the manufacturing process within a fraction of a second and it builds the benchmark function of the chosen PSO algorithm (NC stands for Numerical Control). The PSO algorithm generates solution candidates and moves these as a particle swarm through the search space to find good solutions. The potential solution parameters represent workpiece positions in the workspace of the virtual tooling machine. With support from the NC interpreter the whole search space can be screened in terms of run time (see Figure 1, (1) and (2)). In order to achieve a rapid run time, the NC interpreter ignores collisions during its calculation. To test the validity of solutions extensive virtual machine runs are required. The K-means as well as adapted binary search algorithm are used to decrease the number of computations required to find valid solutions. The K-means-PSO combination clusters the calculated solutions according to their Euclidean distances within the search space (see Figure 1, No. (4)). The number of clusters depends on the available resources for virtual tooling instances.

The solutions within each cluster are ordered by their fitness value afterwards. The adapted binary search approach splits the data sample, using areas instead of traditional clustering. Therefore the search space gets split in areas of equal size, with the amount of areas depending on the available resources. The solutions of each area are sorted according to their fitness value as well.

After splitting the complete sample with one of the two algorithms the candidate with the best fitness is evaluated, in terms of validity, for each cluster. Once a valid solution is found, it is saved and all solutions with lower or equals fitness are being dropped. Every invalid solution gets dropped as well.

These approaches are guaranteed to find the best valid solution due to their completely deterministic behavior, assuming the number of iterations is not limited. The important factor in order to compare these two approaches has to be which approach needs fewer iterations to find this solution. The evaluation of this metric will be discussed in the following sections.

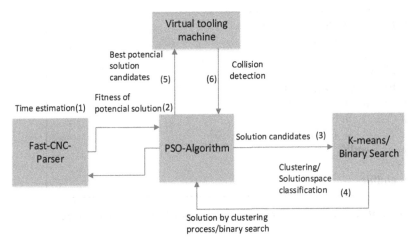

Fig. 1. Overview about the system elements of the PSO extension

3   Use Cases of the Experimental Design

For a successful implementation and comparison of the treated cluster methods in the context of the optimization of setup positions within virtual production, several use cases are developed as experimental design. Figure 2 illustrates four use cases. These are implemented in a self-written use case evaluator as 2D-application. The representation of the system elements are also given in Figure 2. Several positions of workpiece, obstacles and workpiece clamps can be represented alongside the contours.

The tool change point and the tool arm, which is controlled by the NC-program, is also defined. The point of origin, which is also important for the NC-program is defined at the top left corner of the workspace (blues area). Obstacles are also illustrated in red. The blue area as workspace is rasterized into discrete millimeter squares in which each occupied square of the total area is represented by a coordination point.

The experimental design estimates only workpiece positions taking into account of the positions of the clamps and obstacles. Target geometry and material removal are ignored. Thus, an evaluation of positions is possible and the PSO extensions can be compared.

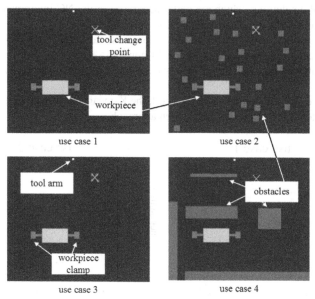

Fig. 2. Use cases overview

4   Results and Discussion of the Findings

The four use cases introduced in Section three (see Figure 2) have been tested with the approach discussed in Section two. In Figure 3, the results for every use case are shown. The X-Axis represents the amount of parallel available virtual machine runs, denoted by k, whereas the Y-Axis shows the iterations needed to find the optimal valid solution of the given sample. The research results are shown in standard box and whisker-plot format, containing minimum, first quantile, median, third quantile and maximum.

The K-means results are visualized in blue and the adapted binary search in red. For every use case the PSO generated 50 independent samples of different sizes depending on the convergence behavior.

In use case 1 the third quartile equals zero for every amount of k, independent of the used algorithm. This is caused by the structure of the use case, which does not have any invalid solutions around the tool change point. This results in the best solutions found by the PSO, being valid and therefore an immediate termination of the algorithm. The few samples using up to 6 iterations contained best solutions, which were right next to the workspace borders, making them invalid.

The number of iterations in use case two is by far the highest. This is resulted the distribution of valid positions, more concrete, most positions around the tool change point are invalid. The clustering approaches are finding the best position, but have to test all invalid solutions, which are better according to the NC interpreter afterwards. This behavior results in a huge amount of iterations.

The use cases 3 and 4 only need few iterations, caused by valid solutions being relatively near to the tool change point. In general a dependency of the amount of needed iterations to the validity of the surrounding of the tool change point can be observed.

Also, in every use case for every tested k the K-means performed either equally as good or better as the adapted binary search approach.

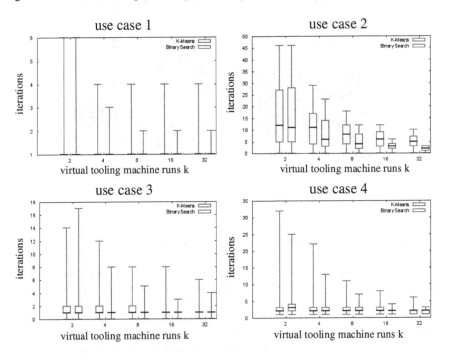

Fig. 3. Experimental results

5   Conclusion and Outlook

This contribution compares two PSO extension approaches which contains the K-means clustering algorithm and the adapted binary search algorithm to provide an intelligent search procedure. The goal is to minimize the required number of collision evaluations in order to find a valid, optimal workpiece position. The experiments are executed by an experimental design instead of a virtual tooling machine. Thus, rapid results are achieved. Simulations have shown the K-means extension to be more efficient in terms of iterations, while finding the same solutions. As this efficiency directly corresponds to the number of virtual machine runs needed for future real world applications, it is the dominant factor in comparing these extensions.

As future work, it is planned to implement the PSO extension in combination with a virtual tooling machine. This way collisions can be detected according to real world machine geometries and parameters. Also material removal and real NC cycles can be simulated. Finally the virtual machine would offer the possibility to determine best positions of workpieces and clamps in a 3D-environment (see Figure 1).

References

1. Alt H: Computational Discrete Mathematics: Advanced Lectures. In: Goos, G., Hartmanis, J., Leeuwen, J. (eds.) LNCE, vol. 2122. Springer, Berlin, Heidelberg, New York (2001)
2. Kennedy, J., Eberhart, R.: Particle Swarm Optimization. In: Proceedings of the 4th International Conference of Neural Networks, vol. 4, pp. 1942 – 1948. IEEE Press (1995).
3. Kennedy, J., R. C. Eberhart and Y. Shi. Swarm Intelligence. pp. 287 – 326. Morgan Kaufmann Publishers, San Francisco. (2001)
4. März, L., Krug W.: Kopplung von Simulation und Optimierung. In März, L., Krug, W., Rose, O. Weigert, G. (Hrsg.). Simulation und Optimierung in Produktion und Logistik, Praxisorientierter Leitfaden mit Fallbeispielen. pp 41 – 45. Heidelberg, Dordrecht, London, New York: Springer (2011).
5. März, L., Krug W.: Simulationsgestützte Optimierung. In März, L., Krug, W., Rose, O. Weigert, G. (Hrsg.). Simulation und Optimierung in Produktion und Logistik, Praxisorientierter Leitfaden mit Fallbeispielen. pp 3 – 12. Heidelberg, Dordrecht, London, New York: Springer (2011).
6. Montgomery, D. C.: Design of Experiments. 8th ed. John Wiley and Sons, Inc. (2013).
7. Reisch, R.-E., Weber, J., Laroque, C., Schröder, C.: Asynchronous Optimization Techniques for Distributed Computing Applications. In: Tolk, A., Padilla, J. J., Jafar, S. (eds.) Proceedings of the 2015 Spring Simulation Multi Conference, 48th Annual Simulation Symposium, vol. 47, no. 2, pp. 49 – 57. IEEE Press, Virginia, Alexandria (2015).
8. Wu, J.: Advances in K-means Clustering: A Data Mining Thinking. Springer, Heidelberg New York Dordrecht, London (2012)

Machine-specific Approach for Automatic Classification of Cutting Process Efficiency

Christian Walther[1,2], Frank Beneke[3], Luise Merbach[3],
Hubertus Siebald[4], Oliver Hensel[4] and Jochen Huster[5]

[1]University of Applied Sciences Schmalkalden, Faculty of Electrical Engineering,
Blechhammer 4-9, D-98574 Schmalkalden, Germany
[2]Fraunhofer IOSB, Advanced System Technology,
Am Vogelherd 50, D-98693 Ilmenau, Germany
[3]University of Applied Sciences Schmalkalden, Faculty of Mechanical Engineering,
Blechhammer 4-9, D-98574 Schmalkalden, Germany
[4]University of Kassel, Agricultural Engineering,
Nordbahnhofstrasse 1a, D-37213 Witzenhausen, Germany
[5]CLAAS Selbstfahrende Erntemaschinen GmbH, Advanced Engineering Electronics,
Münsterstr. 33, 33428 Harsewinkel, Germany

Abstract. The identification of an inefficient cutting process e.g. in self-propelled harvesters is a great challenge for automatic analysis. Machine-specific parameters of the process have to be examined to estimate the efficiency of the cutting process. As a contribution to that problem a simple method for indirect measurement of the efficiency is presented and described in this article.

To establish a general algorithm, the vibration data of a harvesting machine were extracted. The data from two sensors were recorded while gathering whole crop silage and while standing still in operation mode. For every data stream, a spectral analysis and a feature extraction was performed.

For the development of the algorithm, exploration techniques of Machine Learning were implemented. Artificial Neural Networks were optimized using subsets of the recorded data and then applied to the independent validation data to compute the efficiency of the cutting process. The established algorithm is able to identify the process efficiency without using additional machine-specific parameters.

The validation results are presented as confusion matrices for each data set and the case-specific population of the generated Artificial Neural Networks. The described algorithm is able to automatically determine an inefficient and machine-specific cutting process as an additional information using vibration data only.

Keywords: classification, condition monitoring, cutting process, machine, sensors, artificial neural network, vibration measurement

1 Introduction

The development and integration of simple and adaptive algorithms into embedded systems is demanded by further technology evolution [1]. The automatic generation of models for analytic approaches can be performed by evolutionary optimization and further techniques of Machine Learning. As an example a standalone embedded diagnostic system for automatic analysis of cutting processes should be developed. As an essential part of this system the automatic approach of data analysis is presented in this work. An inefficient cutting leads to a poor process quality and to a higher energy consumption for example in self-propelled field harvesters [2]. Additionally a frequently grinding of the chopping blades leads to higher costs and requires more machine downtime [2]. So the cutting process efficiency in self-propelled field harvesters should be analysed to determine the perfect point of grinding. In advantage to [3] it is required that the analysis does not need any additional machine specific information or to know about the type of crop.

2   Material and Methods

Relationships between the measured data and recognized states can be found by strategies of Machine Learning. By using meaningful data and features efficient models of classification can be generated for example by Evolutionary Algorithms. Next to the machine-specific vibration data additionally the telemetry of the machine was available. The telemetry is only used for control purposes and not used by the automatic classification algorithm. The built in accelerometers are normally used to control the positioning of the cutting system after the process of grinding the chopping blades. The grinding process requires the harvester to stand still. But while harvesting the sensors are not needed by the machine. The machine generates vibration during operation. The vibrations seem to be different in relation to the sharpness of the chopping blades and in common to the efficiency of the cutting process [3][4][5]. The cutting process was divided into three different states *inefficient*, *normal* and *efficient*.

2.1   Data Aquisition

The whole data were collected throughout two case studies using a Claas Jaguar 950 as shown in figure 1. The self-propelled harvester was used to gather whole crop silage in the field in the first case A. In the second case B it was in working mode without gathering crop and it was standing still. In a first step in both cases data were collected while using not grinded chopping blades. These data were rated as a *inefficient* cutting process. After the first data recording the harvester runs grinding cycles for the whole chaff cutting assembly. The next data sets were collected after grinding and these data were rated as a *normal* cutting process. Grinding cycles for the whole chaff cutting assembly were performed again. Afterwards the last data sets were recorded and marked as *efficient* cutting process. So independent data sets were generated for evaluation. For every state of the cutting process in case A and B approximately 180 to 220 epochs of data were recorded. In the first case one grinding cycle was performed between the states of the process and in the second case two cycles. The vibration data were recorded at a sampling rate of 51.2 kHz with a *NI USB-4431* device at two different positions near the rotating chopping blades. By using two accelerometers two different data channels were observed. The features were extracted offline. The feature base was defined by several variants of the Spectral Edge Frequency (SEF). The SEF is mainly used in monitoring and classification of electroencephalographic patterns in anaesthesia [6][7]. The original signals and feature values were directly used for analysis. The data sets were divided into epochs for training, test and validation of the Artificial Neural Networks (ANN).

Fig. 1. Self-propelled field harvester (left) and its chaff cutting assembly with 2x12 chopping blades (right). Two built in accelerometers are used to evaluate the cutting process for two channels.

2.2 Requirements

The long term aim is to develop an independent embedded system which is automatically self learning using the described algorithm in this article. To ensure the feasibility in a first step a software system was developed to analyse the vibration data and the cutting process. The first results concerning this work are described in this article. The software should be able to generate process-specific models within a *Training* procedure. In addition the generated models should be used to classify data within an *Application* procedure to confirm the quality. The number of states of the cutting process depends on the classification task. The optimization of the models is performed automatically.

2.3 Software Development Process

To establish an automatic optimization and classification approach a standalone software system named *EMiL-Analyzer* was developed. The software system was new programmed and based upon the experiences in software programming for the systems partially described in [6] and [7]. The software development was bound locally. The Evolutionary Development Model for Software was chosen to react to changes of requirements rapidly [8][9]. Additionally the software was programmed using components for later reuse and maintenance. The main components consist of data management, spectral analysis, feature extraction, evolutionary optimization, artificial neural networks and the user interface. The Software *EMiL-Analyzer* is still working in offline mode and provides the analysation and modeling algorithm as described below. An automatic online mode for classification tasks was prepared.

2.4   Evolution and Modelling

The relationship between the recorded data and the states of the cutting process is modelled using Artificial Neural Networks (ANN). The training of these models is performed with error backpropagation. The structure of the ANN and parameters of the backpropagation algorithm are optimised using an Evolutionary Algorithm. Multilayer Perceptrons with two hidden layers and sigmoid activation functions were optimized. Because the number of states could be changed in subsequent studies, neural networks were chosen to optimize. The implemented Evolutionary Algorithm uses a elite population of individuals. The size of this population was 10, the number of child individuals was 5. The number of generations for all cases was 40. The rate of recombination for cross over of chromosomes was 0.7, the rate of mutation of a gene position was 0.3. The mutation of genes is performed as described in [7]. The procedure of automatic generation of models, i.e. ANN for classification of efficient cutting processes, is called *Training* in this work. The later use of the models is called *Application*. Both procedures are explained in the following and illustrated in figure 2.

The data sets, which were recorded throughout the two different cases, were divided into subsets for training, test and validation. The subsets of training and test were used for the procedure of *Training* of the ANN. To calculate the performance as presented in tables 1(a-d) the independent subset of validation was used. In the following the procedure of *Training* is explained:

1. Initialisation of the elite and the child population with a given number of individuals and ANN (one ANN corresponds to one individual). The structure of the ANN ist randomly initialized.
2. Training of the ANN with error backpropagation using the vibration training data and the corresponding process classification.
3. Evaluation of the ANN using the vibration test data and the process classification. Calculation of $q_{1,i}, q_{2,i}, q_{3,i}$ and $q_{4,i}$ as described in the next paragraph for each ANN i.
4. Calculation of the fitness f_i of the individuals using the quality of the ANN as a weighted sum: $f_i(q_{1,i}, q_{2,i}, q_{3,i}, q_{4,i}) = 0.3*q_{1,i}+0.2*q_{2,i}+0.2*q_{3,i}+0.3*q_{4,i}$ where i indicates the individual and the corresponding ANN.

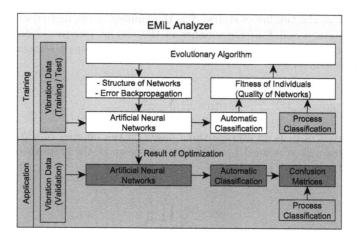

Fig. 2. The developed software *EMiL-Amalyzer* provides a built in optimization procedure *Training* and a procedure of *Application*. Within the *Training* procedure populations of ANN are optimized using an Evolutionary Algorithm. Within the *Application* the resulting ANN are used to classify unknown vibration data as a *inefficient, normal* or *efficient* cutting process. For performance evaluation confusion matrices are calculated.

5. Storage of the best individuals into the new elite population. Generation of new child population using the operators of selection, recombination and mutation. Continue with step 2 until generation limit is reached.

   Finally the procedure of *Application* is described:

1. Evaluation of the validation data for every generated ANN. Within a real application of the system in future this data will be recorded online and rated automatically by the ANN.
2. Calculation of confusion matrices as presented in tables 1(a-d) for validation only.

2.5   Evaluation of the Artificial Neural Networks

The Artificial Neural Networks are evaluated within the evolutionary optimization using the quality criteria which were presented in [6] and [7]. These criteria are used to compute a fitness value for each individual within the evolution. A neural network is an attribute of an individual next to the fitness value. The first criterion evaluates the quality of the absolute accordance over all epochs or datavectors and is defined by

$$q_1 = \frac{\sum_{n=1}^{N} I_n}{\sum_{n=1}^{N} S_n} \tag{1}$$

with N as the number of states, S as the number of the desired data vectors for each state n and I as the number of the correct recognized data vectors for each state n. To evaluate the accordance for every single state n a second criterion is defined by

$$q_2 = \frac{1}{N} \sum_{n=1}^{N} \frac{I_n}{S_n}. \tag{2}$$

The result of q_2 is an average value based on the number of states N. The third criterion q_3 is defined by the geometric mean of the state-specific accordance and calculated with

$$q_3 = \sqrt[N]{\prod_{n=1}^{N} \frac{I_n}{S_n}}. \tag{3}$$

At least the criterion q_4 is presenting the worst accordance of all states with

$$q_4 = \min_{n=1,2,\dots,N} \left(\frac{I_n}{S_n} \right). \tag{4}$$

To obtain an individual-specific fitness value, a weighted sum is calculated using the values of the quality for each neural network. The fitness value is connected to an individual, which itself represents the corresponding neural network within the evolutionary optimization process.

3 Results

Optimized populations of Artificial Neural Networks were generated to classify the states of a cutting process. Therefore a standalone software for automatic generation and application of this networks was developed. The optimization of other models of classification is also prepared in this software. An Evolutionary Algorithm was implemented to generate the mathematical connection between the recorded vibration data and the states of a machine-specific cutting process. To validate the generated ANN an independent subset of vibration data was used. The prediction of the states of the cutting processes of this subset shows a very high concordance as presented in the tables 1(a-d).

The general confusion is on a very low level. Additionally in the tables 1(a) and 1(b) the reached performance is very similar for channel 1 and 2. The same result is recognisable in the tables 1(c) and 1(d) for the second case study. The use of two independent channels allows additional control of the validation.

4 Conclusion and Future Works

Next to the obtained objectives several other results can be concluded. It was shown that the feature Spectral Edge Frequency is suitable for machine diagnostics. The SEF is normally used to classifiy patterns of electroencephalograms

(a)

A Channel 1	Automatic Classification		
	inefficient	normal	efficient
Process inefficient	94.4%	0.0%	1.9%
Process normal	0.0%	74.1%	25.9%
Process efficient	0.0%	3.0%	97.0%

(b)

A Channel 2	Automatic Classification		
	inefficient	normal	efficient
Process inefficient	100.0%	0.0%	0.0%
Process normal	1.9%	87.0%	0.0%
Process efficient	0.0%	4.5%	95.5%

(c)

B Channel 1	Automatic Classification		
	inefficient	normal	efficient
Process inefficient	100.0%	0.0%	0.0%
Process normal	0.0%	88.9%	11.1%
Process efficient	0.0%	0.0%	100.0%

(d)

B Channel 2	Automatic Classification		
	inefficient	normal	efficient
Process inefficient	100.0%	0.0%	0.0%
Process normal	0.0%	87.0%	13.0%
Process efficient	0.0%	0.0%	100.0%

Table 1. Median tables of confusion matrices generated by the resulting populations of artificial neural networks. The values of the tables are calculated for each case A and B and for each channel 1 and 2. For each table the performance of 15 ANN was evaluated using the independent validation data.

througout anaesthesia procedures [6][7][10]. Furthermore, very similar accordances for channels 1 and 2 were calculated for each of the cases as presented in tables 1(a-d). The use of two channels allows an additional internal control of the generated calculation results of the algorithm. The presented process of training should also work for different machines. The generated models for classification are machine specific in every case. Throughout this work only one machine was evaluated, so further research should consider different machines and crops. The optimizing evolutionary algorithm should be expanded with the implementation of additional evolutionary selection procedures, other classification models like fuzzy rules and an automatic feature selection. To increase the performance and robustness of the populations of neural networks the approach of cooperation as presented in [6] and [7] should be integrated too. The implementation of Multi-objective Evolutionary Algorithm as presented in [7] should lead to an additional increase of performance. A forecast of a change of the cutting-process efficiency should be added to present a time-based point of grinding. In future research the data recording, data preprocessing and the presented algorithm should be implemented into an embedded system [11][12].

The project was supported by funds of the German Governments Special Purpose Fund held at *Landwirtschaftliche Rentenbank*.

rentenbank

References

1. Arbeitskreis Embedded Architekturen: Eingebettete Systeme der Zukunft - Charakteristika, Schlüsseltechnologien und Forschungsbedarf, BICCnet Garching (2014)
2. Wild, K., Walther, V.: Verschleiß und Schleifen der Messer beim Feldhäcksler - Der richtige Schleifzeitpunkt, Lohnunternehmen, pp. 40-43 (2012)
3. Heinrich, A.: Verfahren und Anordnung zur Bestimmung der Schärfe von Häckselmessern - Method and device to determine the sharpness of chaff-cutting blades. Patent EP1386534B1 (2003)
4. Siebald, H.: Körperschallmessungen Feldhäcksler CLAAS Jaguar 950 - akustische Messerschärfezustandserkennung - Messbericht. University of Kassel, Agricultural Engineering Witzenhausen, Germany (2012)
5. Merbach, L.: Akustische Messerzustandserkennung - Machbarkeitsstudie, Bachelorthesis, University of Applied Sciences Schmalkalden, Germany (2013)
6. Wenzel, A.: Robuste Klassifikation von EEG-Daten durch Neuronale Netze - Untersuchungen am Beispiel der einkanaligen automatischen Schlafstadien- und Narkosetiefenbestimmung. Shaker Verlag, Aachen (2005)
7. Walther, C.: Multikriteriell evolutionär optimierte Anpassung von unscharfen Modellen zur Klassifikation und Vorhersage auf der Basis hirnelektrischer Narkose-Potentiale. Shaker Verlag, Aachen (2012)
8. May, E. L., Zimmer, B. A.: The Evolutionary Development Model for Software, Hewlett Packard-Journal (1996)
9. Gilb, T.: Competitive Engineering: A Handbook for Systems Engineering, Requirements Engineering, and Software Engineering Using Planguage, Butterworth-Heinemann (2005)
10. Walther, C., Baumgart-Schmitt, R., Backhaus, K.: Support Vector Machines and Optimized Neural Networks - Adaptive Tools for Monitoring and Controlling the Depth of Anaesthesia, The 3rd International Conference on Electrical and Control Technologies, Kaunas, Lithuania (2008)
11. Schneider, M., Wenzel, A.: Entwurf eines eingebetteten Diagnosesystems zur Überwachung von Prozessparametern bei Spritzgießen, pp. 91-105, FHS-prints 2/2014 (2014)
12. Fraenzel, N., Weichert, F., Wenzel, A., Ament, C.: A prototyping system for smart wheelchairs, Biomedizinische Technik, Vol. 59, pp. 902-905 (2014)

Meta-analysis of Maintenance Knowledge Assets Towards Predictive Cost Controlling of Cyber Physical Production Systems

Fazel Ansari & Madjid Fathi

Institute of Knowledge-Based Systems & Knowledge Management, Department of Electrical Engineering & Computer Science, University of Siegen
Hölderlinstr. 3, 57076 Siegen
{fazel.ansari,madjid.fathi}@uni-siegen.de

Abstract. Successful transition to Industry 4.0 requires cross domain and interdisciplinary research to develop new models for enhancing data and predictive analytics. Predictive models in particular should be applied to real time and remotely maintenance cost planning, monitoring and controlling of cyber physical production systems (CPPS). This paper presents a knowledge-based model, *Costprove*, discusses its mathematical meta-analysis approach for evidence extraction, and studies its application in the state-of-the-art industry towards its prospective in causality detection and predictive maintenance cost controlling of CPPS.

Keywords: Maintenance, Cost Controlling, Meta-analysis, Knowledge Assets, CPPS.

1 Introduction

Projection of existing maintenance principles and models has been recently in the research focus to address the challenge of maintenance of cyber physical production systems (CPPS) (Ruiz-Arenas, et al., 2014),(Sharma, et al., 2014) and (Niggemann & Lohweg, 2015). The ultimate goal of these research endeavors is to outline the ideas towards developing cyber system-specific maintenance models. Maintenance cost management (MCM), consisting of cost planning, monitoring and controlling, is an essential part of the sustainable and efficient maintenance management system. MCM is determined as a knowledge-centered and experience-driven process where exploiting existing knowledge and generating new knowledge strongly influences every instance of cost planning. The key aspect of MCM is learning from past experiences for continuous improvement of the maintenance cost planning and controlling. Taking into account the evolution of knowledge-based maintenance (Biedermann, 2014), the authors have discussed the meta-analysis of knowledge assets for MCM of production systems in their former publications cf. (Ansari, et al., 2014),(Ansari, 2014) and

(Dienst, et al., 2015). The *Costprove*[1] model has been therefore constituted to deploy knowledge assets of maintenance for improving MCM through continuous learning from the past events (former planning periods) and experiences (Ansari, 2014). In each planning period (*p*), the chief maintenance officer (CMO) defines cost figures (i.e. planned, unplanned or total cost) and correspondingly the operational factors such as number of maintenance activities. Attainment of the predefined goals is monitored and controlled during and after the planning period respectively. To identify figures of the forthcoming period (*p+1*), the CMO studies records of the past planning period (*p*), and examines the deviations. The result is the recognition or discovery of certain facts with regard to the state of MCM. The cost planning, monitoring and controlling process, thus exploit existing knowledge, and generates evidences (i.e. facts and/or artifacts) that need to be further explored. Anticipating the fundamental challenge of MCM in CPPS, this paper introduces the *Costprove* and its meta-analytical approach (cf. Section 2), studies its sample application (cf. Section 3) and finally in Section 4 outlines its potentials for assisting the CMO in causality detection based on the extraction of evidences in the context of CPPS (cf. Figure 1).

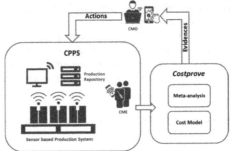

Fig. 1. Integration of *Costprove* for MCM of CPPS

2   *Costprove* Model

2.1   Qualitative Modeling

The *Costprove* model (cf. Figure 2) consists of a meta-level (A-B-C), and basic or object-level (D-E-F). The process flow consisting of the blocks D-E-F is a closed loop, in that maintenance operation is planned and conducted. Sub-loop D-E is to monitor the maintenance process and related activities. The attainment of the planned goals is monitored and reported to meta-level. The chief maintenance engineer (CME) uses standard mathematical models/KPIs (cf. Component E) to plan and monitor maintenance operations and related expenditures. These mathematical models or KPI can differ based on the requirements for each special use-case scenario. He/she is, therefore, assisted to plan, conduct or predict maintenance schedules, programs and events in the plant (cf. component F). Ultimately the CME can estimate required re-

---

[1] *Costprove* stands for two-level maintenance cost controlling system for continuous learning and improvement in MCM.

sources, budget and planned costs for the financial calendar (such as forthcoming months, or year). Also the CME can optimize workload and interval of tasks to comply with the given budget (cf. component C). Only those activities which are successfully tested in the sub-loop will be realized, and, of the realized ones, only the approved activities are documented and reported by the CME. Thus the object-level includes an internal step in learning from the initial estimations and improving the planning and monitoring of maintenance.

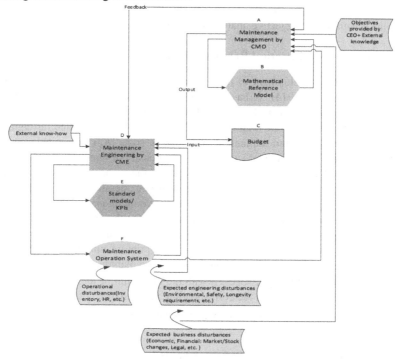

Fig. 2. Qualitative modeling of *Costprove*

At the meta-level, the CMO is responsible of controlling whether the planned operational objectives are achieved, based on estimated (planned) cost and available budget. The process of controlling is to (1) assess and examine the desirability of the current (actual) state, (2) detect improvement potentials, and (3) (re)formulate strategies for the next planning period (e.g. financial year). The CMO receives feedback via multiple channels. He/she monitors the activities of the CME and has direct access to operational information. The CMO has to communicate with senior managers (CEO) to assure accomplishment of business (production economy) objectives. Therefore, he/she should consider multiple factors in making decisions. The CMO bridges the "planning-monitoring-controlling" gap, by indicating deficiencies and improvement potentials, (re)design of strategies, and refining the budget estimates. Thereby the CMO controls and supervises the proper function of the D-E-F loop, and considers organizational preferences and expected business disturbances (e.g. increase of wages due to economic crises, financial or stock market changes, legal issues). To make a decision, the CMO needs to identify assumptions, define certain alternatives, estimate risks and select the best solution (cf. sub-loop A-B) by means of the mathematical reference model (cf. Component B). The CMO can distinguish and describe the problem, but to (re)design strategies he/she crucially should be aware of historical data and collected information within maintenance activities. The CMO uses the maintenance database to review the documentations regarding the current state of operation and expenditures. He/she, therefore, may effectively use his/her domain expertise, and combine it with related objective analysis of maintenance records.

2.2   *Costprove* Mathematical Reference Model

The design and development of the *Costprove* mathematical reference model are discussed in (Ansari, 2014). It has been inspired by the incomplete mathematical model of (Hahn & Laßman, 1993) which qualitatively represents the cost model (cost/benefit ratio), where the total maintenance cost (K_I) is interpreted through indication of planned (K_{VI}) and unplanned maintenance costs (K_S) for a single machine as: $K_I = K_{VI} + K_S$. The argument (n) implies the intensity of maintenance activities in a certain period that on the one hand influences on the planned maintenance operations cost, and on the other, the unplanned maintenance operations cost in that period. In other words, (n) indicates proceeding of preventive maintenance activities to optimize the planned and unplanned maintenance cost. Since the unplanned cost function is non-linear and planned cost function is linear, the summation curve is expected to have a single (global) minimum that reveals the optimum number of maintenance activities (n_{opt}), e.g. for a single machine in the certain planning period (e.g. per quarter). This optimum is directly associated with the minimum of total maintenance cost $(K_{I_{min}})$ of the single machine in the corresponding maintenance costing period.

In order to develop the *Costprove* mathematical reference model two case distinctions are considered as:

- *Case I* is to formulate a cost/activity ratio and derive the optimum number of maintenance activities, and minimum of total maintenance cost. It is developed for *a single production machine*, and only considers preventive maintenance (PM) costs (cf. Table 1). *Case I* is extended to employ a method for calculating total maintenance cost, based on time-dependent/independent cost items (including PM and corrective maintenance (CM) costs) of maintenance for *disassociated (non-associated) single production machines* (cf. Table 1).
- *Case II* is to verify the outcomes of *Case I* by examining whether the net total cost (excluding unplanned cost) is less or more than the assigned budget. This may lead to further investigations for identifying the planned/unplanned cost raised due to the association of the machines, or internal/external disturbances. They directly have an effect on overall maintenance costing. In other words, it is to broaden the assumption of Case *I* and consider *the effect of shortage situation* (cf. Table 1).

Table 1. The developed formulations of *Costprove* (*Case I* and *Case II*) (Ansari, 2014)

		Expression	Formulation
Case I	**Single machine**	Modeling of unplanned cost with exponential function	$n_{opt} = -\frac{1}{a_S^*} \cdot Ln\left(\frac{a_{VI}}{a_S \cdot a_S^*}\right)$ $K_{I_{min}} = \frac{a_{VI}}{a_S^*} \cdot \left(1 - Ln\left(\frac{a_{VI}}{a_S \cdot a_S^*}\right)\right) + (b_{VI} + b_S)$ $0 < \left(\frac{a_{VI}}{a_S \cdot a_S^*}\right) < 1$
		Modeling of unplanned cost with power function	$n_{opt} = \left(\frac{r \cdot a_S}{a_{VI}}\right)^{\frac{1}{r+1}}$ $K_{I_{min}} = \left(a_{VI} \cdot \left(\frac{r \cdot a_S}{a_{VI}}\right)^{\frac{1}{r+1}} + a_S \cdot \left(\frac{a_{VI}}{r \cdot a_S}\right)^{\frac{r}{r+1}}\right) + (b_{VI} + b_S)$ $n \neq 0$ • The subscripts for total (I), planned (VI) and unplanned (S) costs.
	Disassociated Single machines	Total maintenance cost	$\underline{K}_I = \underline{t} \cdot \underline{C} + \underline{K}_e$ $\begin{bmatrix} K_{I,11} & \cdots & K_{I,1\rho} \\ \vdots & \ddots & \vdots \\ K_{I,\mu1} & \cdots & K_{I,\mu\rho} \end{bmatrix} =$ $\begin{bmatrix} t_{11} & \cdots & t_{1\sigma} \\ \vdots & \ddots & \vdots \\ t_{\mu1} & \cdots & t_{\mu\sigma} \end{bmatrix} \cdot \begin{bmatrix} C_{11} & \cdots & C_{1\rho} \\ \vdots & \ddots & \vdots \\ C_{\sigma1} & \cdots & C_{\sigma\rho} \end{bmatrix} + \begin{bmatrix} K_{e,11} & \cdots & K_{e,1\rho} \\ \vdots & \ddots & \vdots \\ K_{e,\mu1} & \cdots & K_{e,\mu\rho} \end{bmatrix}$ $\underline{K}_I = [\underline{t}_{Pl} \cdot \underline{C}_{Pl} + \underline{t}_{Unpl} \cdot \underline{C}_{Unpl}] + [\underline{K}_{e_{Pl}} + \underline{K}_{e_{Unpl}}]$ • (Pl) and ($Unpl$) are the subscripts of the matrices for indication of the planned and unplanned classification of cost, time and cost rate of maintenance activities.

Parameter	Expression	Dimension
$K_{I,\mu\rho}$	Total maintenance cost (i.e. identifiable planned or unplanned cost) for the machine $\mu = (1,...,m)$ within the maintenance plan $\rho = (1,...,p)$	*Cost unit*
$t_{\mu\sigma}$	Time-dependent parameter for maintenance of machine $\mu = (1,...,m)$ corresponding to maintenance activity $\sigma = (1,...,s)$	*Time*
$C_{\sigma\rho}$	Cost rate of maintenance activity $\sigma = (1,...,s)$ over maintenance plan $\rho = (1,...,p)$	*Cost/Time*
$K_{e,\mu\rho}$	The residual maintenance cost for machine $\mu = (1,...,m)$ within the maintenance plan $\rho = (1,...,p)$	*Cost unit*
m	Number of machines	---
p	Number of maintenance plans	---
s	Maintenance activity	---

		Expression	Formulation
Case II	**Associated machines**	Shortage situation	$TMB < \sum_{i=1}^{m} (K_{I_{min}}^i - TUC_i)$ $\underline{TUC} = \underline{t}_{Unpl} \cdot \underline{C}_{Unpl} + \underline{K}_{e_{Unpl}}$ • TMB stands for the *"Total Maintenance Budget"* assigned to the CMO. • $K_{I_{min}}^i$ is the minimum of total maintenance cost of a machine i (cf. matrix representation of *Case I* where $m = i$). • TUC_i denominates the *"Total Unplanned Cost"* of the machine (i) in the certain planning period (p).

Case I is developed within a heuristic procedure whereby the CMO identifies K_{VI}, based on existing documentation (e.g. historical data or technical reports) as well as his/her domain expertise (cf. Figure 2). Here the quality of accumulated maintenance records (in maintenance databases) and competence of the CMO are significant. In contrast to planned cost, determining unplanned cost (K_s) is difficult in an empirical way and requires the investigation of downtime consequences (e.g. for associated machines, processes, human resources, logistics and business process). K_s , therefore, is an opportunity cost which provides a possibility for improving MCM. Using the *Case I*, the CMO needs planned or unplanned costs to find the minimum of total cost and optimal number of maintenance activities through identifying the planned and unplanned parameters ($a_{VI}, b_{VI}, a_s, a_s^*, b_s$ and r). The parameters are defined based on linear modeling of the planned cost and unplanned costs with exponential or power functions (cf. Table 1). The extension of *Case I* also supports the CMO for improving the modeling of the maintenance cost by scaling up the number of machines (production lines), and considering different maintenance planning periods including past and forthcoming events (cf. Table 1).

Table 2. Guideline to identify tasks of planning-monitoring-controlling in *Costprove*

	Planning (in period p with figures for period $p + 1$)	Monitoring (in period $p + 1$ with figures for period $p + 1$)	Controlling (in period $p + 1$ with figures for periods t and $p + 1$)
Planned maintenance cost	Defining the parameters as targets, based on experience, historical data, and actual goals of the company	Achieving the real parameters from the period $p + 1$	Comparison of planned and realized figures, analyzing the deviations **Outcome:** Information to improve efficiency of maintenance and efficiency of planning
Unplanned maintenance cost	Formulation of the cost function, including defining the parameters as a prognosis, based on assumptions, historical data, etc.	Achieving the real parameters from the period $p + 1$	Comparison of predicted and realized figures, analyzing the deviations **Outcome:** Information to improve the forecast of the parameters of unplanned cost in period $p + 2$
Total maintenance cost	Calculating the sum of planned and unplanned costs	Calculating the sum of planned and unplanned costs	See above
Number of maintenance activities	Calculating optimum by minimizing total cost	Ascertaining the actual number	Analyzing the deviations, improvement of the planning process in period $p + 2$

In practice, the CMO should assign a certain budget for maintenance activities of (a set of) machines in a certain period (p). He/she essentially should know how to reach the minimum of total maintenance cost through the optimal distribution of maintenance activities and estimation of the association effect of the machines. Coincidentally, the CMO needs to prevent exceeding the budget limit (i.e. shortage situation). Therefore the budget is known and distribution of maintenance activities with respect to the type of association is unknown. In addition to the effect of association, internal and external disturbances may raise extra costs. Examples are operational disturbances like sudden lack of physical or human resources, and also business disturbances such as economic crises and changes. A figure of merit is required here to examine: (i) in the planning phase, whether the planned cost is sufficient, and (ii) in the controlling phase, whether the planned cost was significant. This process ultimately leads to learning from past experiences within the planning-monitoring-controlling of MCM. A shortage situation occurs when the total budget for a set of machines (including serial or parallel association), in the presence or absence of disturbances, is less than the net sum of total maintenance cost. The figure of merit for the shortage situation is represented in Table 1. The tasks of planning-monitoring-controlling in the context of *Costprove* are described and detailed in Table 2. It elaborates the major tasks in each phase (i.e. planning, monitoring and controlling) in relation to the parameters of the mathematical model.

3   Application Study of *Costprove*

The *Costprove* model is studied for a use-case scenario of a vessel manufacturing company (VMC) which produces pressure and storage vessels and vessels' parts. The production type is an engineer-to-order. The case study has been implemented for cost controlling of two computerized numerical control (CNC) machines where both machines were identical; however, CNC_1 reached 60% of its lifetime while CNC_2 passed 30%. The analysis has been applied in the course of annual cost planning periods of the machines particularly for two planning periods. The duration of each planning period was six months. The procedure of the analysis has been defined in certain steps as:

1. CMO provides the actual number of PM activities and associated total cost over the planning periods as well as his/her estimations for optimal values.
2. CMO uses the mathematical model (based on the formulas given in Table 1 – *Case I*) and starts initiating the calculations on a heuristic manner based on his/her domain expertise.
3. CMO validates the calculations based on the procedure described in Table 1 – *Case II* (i.e. criteria for shortage situation).

Table 3. Result of the case study of *Costprove* in VMC

Empirical data recorded from the actual situation in planning period p				Optimal situation estimated by CMO without *Costprove* model for planning period $p+1$		Optimal situation estimated by CMO with *Costprove* model for planning period $p+1$		Opportunity cost (difference between optimal estimations)	
Maintenance plan	Type of machine	Number of actual PM	Actual total cost	Optimum number of PM	Optimum total cost	Estimated PM	Estimated cost	Saved cost	Percentage
1	CNC_1	67	5834.00	62	5500.00	58	5389.00	111.00	~2.02 %
1	CNC_2	54	2510.00	52	2300.00	48	2134.50	165.50	~7.20 %
2	CNC_1	75	6200.00	69	5800.00	63	5450.00	350.00	~6.04 %
2	CNC_2	58	2590.00	52	2450.00	49	2185.50	264.50	~10.79 %
Total			17134.00		16050.00		15159.00	891.00	~5.55 %

Besides, CMO used two propositions: (i) Maximum budget assigned to PM is 8,000 for both machines in each planning period, and (ii) Maximum number of PM activities should not exceed 70 per planning period for each machine.

The results of the analysis are presented in Table 3. It consist of four columns representing (1) empirical data gathered from reports of the engineers and operators, (2) the optimal situation estimated by the CMO without use of *Costprove*, (3) the optimal situation estimated by the CMO with the use of the *Costprove* and (4) the detected opportunity cost. Notably, the cost quantities are disassociated with standard units of measurement. Consequently, the results reveal the significant opportunity cost through comparing the optimal estimations implemented by CMO with and without use of the *Costprove* model (cf. Table 3).

4   Conclusion and Future Research Agenda

Deployment of CPPS raises several challenges for industries with regard to potential changes in processes, systems and human-based skills and competencies. This article introduces the *Costprove* model and its mathematical modeling approach for planning, monitoring and controlling of the maintenance cost figures. Development of the *Costprove* is underlying a premise that planning-monitoring-controlling is implemented by the interaction of humans such as the CMO. The MCM is, thereby, controlled semi-automatically.

The future research agenda is firstly to improve the accuracy and reliability of the estimations of the *Costprove* model through analysis of the probability of failures in association with planned and unplanned maintenance costs, which has been studied by the authors in (Dienst, et al., 2015), is secondly to implement and integrate an automatic learning mechanism e.g. by means of deep learning, and is thirdly to combine analysis of structured and unstructured knowledge assets i.e. combining mathematical and textual meta-analysis (see (Ansari, et al., 2014)).

By enhancing the *Costprove* the maintenance management system of CPPS will be supported through learning from multi-modal data reflecting the state, failures and deficiencies of the past planning period, and the CMO will gain proactive recommendation assistance to apply required changes in the running or forthcoming planning period. Hence, advancement and evolution of the *Costprove* model will supply the complex demands of CPPS to a robust cost-analytics and predictive model.

References

Ansari, F., 2014. *Meta-analysis of knowledge assets for continuous improvement of maintenance cost controlling.* Dissertation, University of Siegen, Germany.

Ansari, F., Uhr, P. & Fathi, M., 2014. Textual Meta-analysis of Maintenance Management's Knowledge Assets. *International Journal of Services, Economics and Management.* Inderscience Enterprises Ltd., pp. 14-37.

Biedermann, H., Ed., 2014. *Instandhaltung im Wandel (Maintenace in Transition).* TÜV Rheinland Group, Cologne, Germany.

Dienst, S., Ansari, F. & Fathi, M., 2015. Integrated system for analyzing maintenance records in product improvement. *The International Journal of Advanced Manufacturing Technology,* 76 (1-4), Springer, pp. 545-564.

Hahn, D. & Laßman, G., 1993. *Produktionswirtschaft-Controlling industrieller Produktion (Production Economy-Controlling Industrial Production).* Heildelberg: Physica-Verlag, p. 353.

Niggemann, O. & Lohweg, V., 2015. *On the Diagnosis of Cyber-Physical Production Systems: State-of-the-Art and Research Agenda.* Austin, Texas, USA, Association for the Advancement of Artificial Intelligence.

Ruiz-Arenas, S., Horváth, I., Mejía-Gutiérrez, R. & Opiyo, E., 2014. Towards the Maintenance Principles of Cyber-Physical Systems. *Journal of Mechanical Engineering,* 60 (12), pp. 815-831.

Sharma, A.B., Ivancic, F., Niculescu-Mizil, A., Chen, H. & Jiang, G., 2014. Modeling and Analytics for Cyber-Physical Systems in the Age of Big Data. *ACM SIGMETRICS Performance Evaluation Review,* 41 (4), pp. 74-77.

Towards Autonomously Navigating and Cooperating Vehicles in Cyber-Physical Production Systems

Adrian Böckenkamp[1], Frank Weichert[1], Jonas Stenzel[2], and Dennis Lünsch[2]

[1] Technical University of Dortmund, Department of Computer Science VII,
Otto-Hahn-Str. 16, 44227 Dortmund, Germany,
[2] Fraunhofer Institute of Material Flow and Logistics,
Joseph-von-Fraunhofer-Str. 2-4, 44227 Dortmund, Germany,

Abstract. This paper presents a (ROS-based) framework for the development and assessment of (decentralized) multi-robot coordination strategies for Cyber-Physical Production Systems (CPPS) taking into account practical issues like network delays, localization inaccuracies, and availability of embedded computational power. It constitutes the base for (a) investigating the beneficial level of (de-) centrality within Automated Guided Vehicle-based CPPS, and (b) finding adequate concepts for navigation and collision handling by means of behavior-, negotiation- and rule-based strategies for resolving or proactively avoiding multi-robot path planning conflicts. Applying these concepts in industrial production is assumed to increase flexibility and fault-tolerance, e.g., with respect to machine failures or delivery delays at the shopfloor level.

Keywords: decentralized coordination, interoperability, Cyber-Physical Production Systems (CPPS), autonomous navigation, Automated Guided Vehicles (AGV), Robot Operating System (ROS)

1 Introduction

Fundamental concepts of Cyber-Physical Systems (CPS) characterized by the idea of Industry 4.0 are, among others, flexibility, interoperability, and decentralization—specifically by means of intercommunication [6]. Production systems based on these concepts take advantage of the ability to handle mass customization (up to lot size 1), decreased vulnerability to (machine) failures, and the flexibility to adapt to varying selling conditions. Essential elements of Cyber-Physical Production Systems (CPPS) are autonomously interacting units equipped with sensors and actuators to interact with their environment and to allow them to be intelligent, cooperative, and communicative.

We are facing a versatile and integrated approach of localization, navigation, and the coordination of the transport vehicles on the shop floor by combining concepts like tight intercommunication, decentralized cooperation, various collision handling strategies, and local decisions based on observations of the environment. Status reporting, monitoring and self-diagnosis are additional important

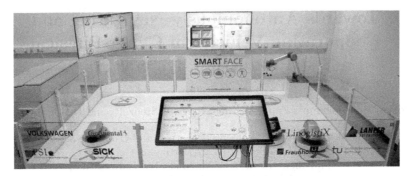

Fig. 1: Initial prototype of our concepts (presented at the Hanover Fair 2015)

features to support the understanding and (inter-) operability of the system. Some of the involved concepts rest upon recent research results, for example, global path planning with arbitrary clearance [7] and Monte Carlo localization [12]. The presented solution is highly modularized to support performance (on embedded devices) and extensions as well as to facilitate scalability to larger and varying setups. If necessary and possible, modules are run on a server back-end (adaptive offloading) to account for missing computational power. Practical issues like communication delays due to wireless connections, limited processing power, localization inaccuracies, and machine failures are taken into consideration as well.

The herein presented approach has several advantages related to CPPS: First, it reduces the system complexity in higher abstraction levels by shifting them to the transport vehicles. Second, it allows them to react on and solve failures on the shop floor on their own which increases flexibility and operability [2]. Third, the system may easily be extended by simply adding new transport vehicles which accounts for scalability.

With the focus on intralogistic material flow handling in industrial production systems, this paper considers the use case of vehicle assembly in the automotive industry. Specifically given the continuously growing trend of electric cars, this use case may exploit all of the previously mentioned advantages of CPPS. For example, a malfunction in a single machine of an assembly *line* causes the whole production process to halt. In contrast to a decentralized CPPS, transport vehicles may ignore a faulty station in favor of selecting another one of the same kind. These aspects are addressed in an interdisciplinary project (SMART FACE), a research endeavor focusing on decentralized concepts for production systems. To overcome the overall problem complexity, different levels of abstractions have been identified to finally approach a demonstrator which incorporates as many of the aforementioned concepts as possible. On the highest level, it aims at implementing the decentralized production concepts in production areas of our industrial project partners. Therefore, a demonstrator is currently being developed at the Fraunhofer IML which consists of an industrial robot, Automated

Guided Vehicles (AGVs), a picking station, and an automated rack. In addition to that, a miniaturized version of the latter has been developed which constitutes the lowest abstraction level, see Fig. 1. The miniaturized version allows us to verify the scalability of our concepts as well as to improve demonstrability and practicability. The approach of this paper was implemented using this setup.

This paper is structured as follows: Sect. 2 describes State-of-the-Art concepts and compares them to our approach. Sect. 3 details the components and their interaction of our ROS-based framework. In Sections 4 and 5, the employed path planning and collision handling concepts are presented. Finally, Sect. 6 summarizes the core ideas and concludes with an outlook regarding further improvements and ongoing work.

2 State-Of-The-Art

Decentralized coordination of AGV fleets is a well-known topic in the literature [3, 9, 13]. Generally, three different types of approaches can be distinguished [3] and we are aiming at a combination of them:

1. Spatial or temporal resource allocation: blocks or segments are blocked before an agent starts off. This decreases efficiency because blocked areas cannot be used by other agents. A more fine-grained level of resource allocation increases the complexity of resource divisions and scheduling considerably.
2. Coordination by agreement and negotiation: agents negotiate precedence and/or priority according to a common protocol. This approach demands extensive information exchange over network and, thus, requires a network connection between the robots. Moreover, if collision avoidance is carried out by means of negotiation, collisions may even occur in case of communication delays.
3. Rule-based approaches: a manually specified or built-in set of rules determines the behavior of an agent. This approach requires fine-tuning the rules and considerable effort to adapt the system to varying conditions.

Digani et al. presented an approach similar to the underlying idea of this paper [3]; it can be associated with type 2. They suggested a concept where the shopfloor is represented in a two-layer fashion and path planning is executed on a two-step basis which results in sub-optimal paths. AGVs are only modeled as "triangles" so that their footprint is disregarded during path planning. In addition, the "topological layer" seems to be composed of (inflexible) rectangular blocks only. Results have only been obtained by simulation whereby neither communication delays nor localization inaccuracies have been considered.

Our approach seizes on ideas of the Any-Com approach suggested by Otte and Correll [9]. They describe a class of algorithms that exploit a network connection between robots (e. g., for collaborative path planning) if available but do not fail if some of the packet transmissions fail. This is an important design property to provide algorithms that still work in practice. In addition, they promote that algorithms can benefit from sharing intermediate data which gets updated

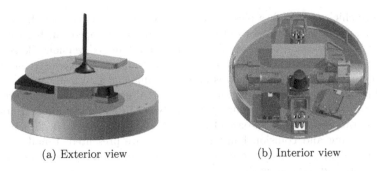

(a) Exterior view   (b) Interior view

Fig. 2: Drawings of the schematic buildup and hardware components (tires, motors, laser scanner, antenna, embedded computing device(s), batteries, and body) of the deployed vehicles

periodically. However, in contrast to our approach, we employ TCP/IP connections instead of UDP to ensure that packets will always arrive at the receiver (maybe with a delay) and, thus, we allow arbitrary delays instead of packet loss. This agrees with the idea of Otte and Correll that "important information is rebroadcast until it is outdated and replaced" [9].

Finally, it should be noted that industrial acceptance can be a problem and may even be an argument against decentralization [1]. Our solution involves detailed status reporting and monitoring to support industrial acceptance and create a deeper understanding of what is currently happening. Also note that we are interested in determining the appropriate level of decentrality (in terms of efficiency) as a function of the system design. For instance, communicating (local) blockings, recurring delays or new views of static obstacles (e. g., dropped objects) to a central server presumably allows the fleet of AGVs to behave more efficiently.

3   System Architecture

This section is used to describe the system architecture, both in terms of hard- and software. A CPPS is typically composed of many parts, suchs as articulated robots, transport vehicles, assembly stations, human workers, and so forth. All parts that are able to communicate with their environment are henceforth called *active agents*. Human workers may in principle be active agents as well because they can be equipped with additional sensors and wearables to track and integrate them in the CPPS. This paper primary focuses on the transport vehicles but the presented concepts may be integrated in the other aforementioned parts, too.

Figure 2 depicts the hardware components of the vehicles (6 in total) used in our setup. Among the typical hardware components like tires, brushless DC motors, the battery, and the body, it is equipped with a 2D laser scanner (LiDAR) of type SICK TiM 561 and an extendable number of embedded computing

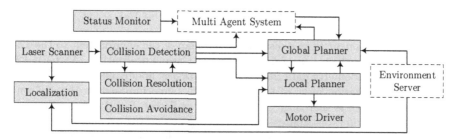

Fig. 3: Simplified view of our software architecture based on the Robot Operation System (ROS) [11]: gray nodes are executed on the agent while white nodes are driven by the server back-end

devices. The latter are connected by a WLAN bridge on the vehicle which, on the one hand, connects the entire vehicle wirelessly (IEEE 802.11, 5 GHz) and, on the other hand, provides Ethernet ports to extend the computing power by adding more embedded devices and connect them locally via Ethernet. This way, a vehicle forms a dedicated and extendable "cluster" of embedded ARM cores which perfectly fits our idea of extensive modularization by means of ROS nodes, described next.

The software components based on the open source Robot Operating System (ROS) [11] are shown in Fig. 3. The *Multi Agent System* (MAS) and the *Environment Server* (dashed borders) are executed on a server back-end while all remaining components (solid borders) are executed on the vehicles individually. The MAS is responsible for managing the active agents in the system and solving the (online) allocation task, i.e., it assigns idle vehicles a new task. New tasks (transport orders) are generated uniformly with parameterizable timing constraints. The allocation is currently performed by maximizing the score function

$$\kappa(v_i) := a_1 \cdot b_i + a_2 \cdot \frac{d_i}{d_{\max}} \in [0,1], \quad \text{with} \quad a_1 + a_2 = 1 \tag{1}$$

among all idle vehicles v_i whereby $b_i \in [0,1]$ denotes the battery level of v_i, d_i is the (Euclidean) distance from v_i to the goal and $d_{\max}$ is the maximum distance. The a_i are weights which allows to customize the selection of the MAS (we used $a_1 = 0.2$ and $a_2 = 0.8$). The MAS manages a list of active agents which is updated periodically by the agent status (a ROS message), sent by the *Status Monitor* node on each agent. Since each agent status contains a time stamp, the MAS is able to detect whether there are non-responding agents in the system. When new agents are added to the system, they start sending their status which automatically updates the list of active agents in the MAS. This way, agents can be removed and (re-) added to the system during operation. Additionally, the MAS is able to collect local event information of the shopfloor like blockings / traffic jams, network latencies or localization uncertainty [5] in order to bundle them in the *Environment Server*. This way, other vehicles can access these information as well.

The *Environment Server* distributes the map of the environment to the *Localization* node which runs Adaptive Monte Carlo Localization (AMCL) [12] (supported by odometry data from the *Motor Driver* node). The *Global Planner* requires a navigation mesh of the environment (superimposed on the environment map) which is shared by the *Environment Server*, too. As already stated, the *Status Monitor* distributes the current agent's status over network. Additionally, it monitors other (critical) nodes (e.g., the collision detection) on the agent as a fail-safe system and checks the battery level periodically. The *Status Monitor* on each agent implements a set of ROS services which provide meta information about an agent, e.g., its radius/size or an approximation of the agent's contour (cf. Sect. 5.2). The system thereby accounts for different types and shapes of agents.

Section 4 describes the path planning and navigation (*Global-*, *Local Planner* and *Motor Driver*) and Sect. 5 explains the nodes related to collision handling.

4   Path Planning

Path planning in the system is carried out by two main nodes: the *Global Planner* and the *Local Planner*. Global planning is done using a navigation mesh (navmesh) which represents the topology of the environment. Since we are interested in finding shortest paths between two locations with a given radius so that the agents are actually able to travel along the path (without colliding with their environment), the concepts of finding shortest paths with arbitrary clearance by M. Kallmann [7] are used. This involves computing a Local Clearance Triangulation (LCT) which may even be updated dynamically [8] (e.g., by adding new obstacles). The triangles in the LCT are represented as a graph and the shortest path is computed using the A* algorithm [4] (leading to sub-optimal solutions). The final sequence of points on the path is then generated using the extended Funnel algorithm [7].

In order to allow the *Global Planner* to re-plan paths around obstacles, the *Collision Detection* node provides obstacle contours for detected objects to the global planner. When the MAS sends a new goal request to the global planner, a new path is computed using the current agent's navmesh. The new path is then sent to the local planner which adjusts the velocities of the *Motor Driver* to travel along the path. If the local planner reaches a goal, a message is sent to the global planner which, in turn, informs the MAS. As mentioned in Sect. 3, the (initial) navmesh is received and updated from the *Environment Server* which allows dynamic collaborative map updates.

5   Collision Handling

In this section, the collision handling is described: it currently consists of *Collision Detection* (Sect. 5.1) and *Collision Resolution* (Sect. 5.2) as behavior-based

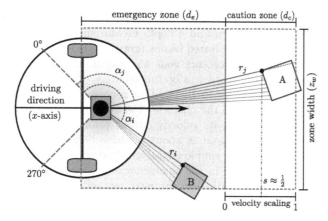

Fig. 4: Scheme of collision detection with adaptive velocity scaling in the caution zone and emergency stop in the emergency zone; the vehicle (black circle) uses the laser scanner (box on x-axis) for generating depth measurements r_i at specific angles α_i. The 2 obstacles "A" and "B" are considered w. r. t. their closed distance to the scanner: "A" would trigger a velocity scaling of about $\frac{1}{2}$ and "B" would initiate an emergency stop

concepts. These concepts will complementary be extended by proactively predicting and avoiding possible collisions by means of broadcasting new paths (Sect. 5.3). Putting this together, we are aiming at developing a framework for analyzing the interaction between behavior-based (reactive) and proactive multi-robot coordination in a decentralized fashion.

The collision detection node continuously senses the environment in front of and alongside the agent using the scan data provided by the *Laser Scanner* node, cf. Fig. 3. It may be extended by additional sensors to allow the agent to detect and classify various types of objects in order to behave differently. The collision resolution node is only triggered when an obstacle has been detected to check whether the obstacle is an active agent and, if applicable, initiate the communication and negotiation. Note that, at this point, network-related problems do not cause any issues since the agents have stopped moving already. When wireless network communication works as expected, all agents share and check their newly computed paths via network so that the probability of possible collisions is reduced to a minimum.

5.1   Collision Detection

Collision detection is the most crucial part for safe operation. Even if collision avoidance is present, one always needs to account for unforeseen events like humans or dropped down objects. The *Collision Detection* node therefore analyzes the laser scan data to detect such upcoming obstacles, as illustrated by Fig. 4.

The laser scanner (box centered on the vehicles x-axis) delivers a set of pairs consisting of an angle α_i and a depth measurement (range) r_i which are computed from the emitted infrared beams (gray lines). The area in front of the vehicle is divided into an emergency zone and a caution zone. Obstacles in the emergency zone cause the agent to stop immediately.

Correspondingly, obstacles in the caution zone reduce the agent's velocity dependent on the distance to the obstacle. The velocity scaling factor s is used by the local planner to scale its velocity output prior to sending them to the motor driver node. If one agent is about to drive behind the other, velocity scaling already implements a simple adaptive "chasing" of the vehicle in front.

If an obstacle is detected in the emergency zone, the collision resolution is triggered which will be described in the next section. However, if an obstacle moves away by itself, collision resolution is aborted and the previous path is continued.

5.2   Collision Resolution

When the collision resolution node is triggered by the collision detection, it receives the scans (α_i, r_i) hitting the obstacle. It also maintains a periodically updated list of agents (like the MAS) using the broadcasted agent status messages. The scans are converted into Cartesian space and the centroid is computed which is then transformed to global (shopfloor) coordinates, approximating the obstacle's position $o = (o_x, o_y)$. The list of active agents then allows to compute the closest agent a_o to o. If the distance of a_o is less than an experimental threshold (accounting for localization inaccuracies), a_o is assumed to be the obstacle. Otherwise, the current agent a_c waits until the obstacle object (not an active agent in the network) moves away and additional processing of the sensor data can be triggered to detect what type of object a_c faces to perform specific tasks/behavior.

Assume now, an agent a_o was found. In this case, a random priority ρ_c is generated and a priority request containing ρ_c is sent to a_o which generates a priority ρ_o, too. If a_o has not (yet) detected a_c as an obstacle, it ignores the request causing a_c to wait (eventually allowing to adaptively drive behind another agent, see Sect. 5.1). If a_o waits anyway, a special (reserved) priority of $\rho_o = 0$ is returned which lets a_c win the priority negotiation. In all cases where a_c wins, the contour approximation of a_o is requested by calling its ROS service so that the global planner is able to compute a new path respecting the obstacle's contour. Alternatively, if a_c loses the priority negotiation, it again waits until the obstacle moves away. If an agent's goal is located at the current position of another (waiting) agent, the latter will always get a higher priority to prevent a deadlock.

5.3   Collision Avoidance

Ideally, neither collision detection (Sect. 5.1) nor resolution (Sect. 5.2) are needed at all because sharing and checking agent paths via network already prevents upcoming collisions before they emerge. However, network-related issues can cause collisions making collision detection and resolution an essential component. For example, if agent a_c communicates its path via network and another agent a_o is currently not reachable, a_o will collide with a_c if a_c's message will not arrive at a_o just before the collision.

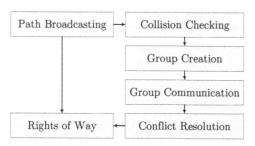 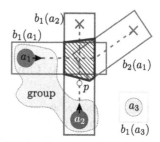

(a) Scheme of collision avoidance by sharing computed paths over the wireless network

(b) Example of collision avoidance via path sharing with three agents

Fig. 5: Steps of collision avoidance (a) and a simplified example (b) of path sharing between agents a_1, a_2, and a_3 whereby a_1 gets the right of way due to negotiations with a_2

Nevertheless, in terms of throughput and efficiency, reducing collisions by proactively avoiding them is the favored solution. Our algorithm (based on ideas of [10]) requires notification messages about newly computed paths (sent by the global planner) and periodic updates about the position, orientation (pose), and current velocity of the active agents in the system. The process of this collision avoidance algorithm is shown in Fig. 5(a). Newly computed paths are broadcasted into the network along with a unique ID (*Path Broadcasting*). Other agents then check received paths by a bounding-box based algorithm (*Collision Checking*). In case of no collisions, the (path, ID)-pair is stored to release unused path segments during further progression of the sending agent. If a collision is detected, the sender agent is notified and the (path, ID)-pair (with the intersecting segments) is stored locally.

Among all agents conflicting with a set of paths, a temporary group (of agents) is created (*Group Creation*). If a group already exists with one of the conflicting paths, all involved agents join the existing group. Creation of groups is priority based: The agent having the smallest number of conflicts with its path (lower workload) is responsible for creating and managing the group. Thus, a group consists of at least 2 and at most N participants (where N is the number of active agents in the system). Within a group, the managing agent determines the right of way at the hypothetical point-of-collision by taking into account the (remaining) travel time and order priority. Each agent in the group is then informed about whether he is allowed to drive or needs to wait at a certain point on its path until an agent of the group releases the appropriate path segments

(*Group Communication* and *Conflict Resolution*). This way, a decentralized negotiation based rights of way principle has been realized. An example of this method with three agents a_1, a_2 and a_3 is visualized in Fig. 5(b). Because the paths of a_1 and a_2 intersect, they form a group (dotted) and negotiate that a_1 has the right of way while a_2 needs to wait at p. a_3 is not part of the group because its bounding box $b_1(a_3)$ is not conflicting.

In terms of offloading, this algorithm may even simply be executed on a server back-end allowing for more complex computations.

6   Discussion & Conclusion

Within this paper, we presented a ROS-based framework for the development and assessment of multi-robot coordination strategies in a decentralized fashion. It consists of several ROS nodes responsible for navigation, collision handling and status monitoring. The MAS constitutes a central point for triggering transport orders and investigating the system status but, in principle, it is not required to operate the system. Previous analysis has shown that employing decentralized concepts in the automotive industry by replacing (parts of) the assembly line can leverage flexibility which is currently not exhausted [2]. This is especially induced by allowing to react to (machine) failures or delays in delivery, i. e., trying to keep the system operational. As a side effect, the complexity in higher abstraction levels decreases.

Given the proposed framework, two areas of research are spawned which will be investigated in the future:

1. The "beneficial" level of (de-) centrality needs to be determined, regarding (a) the computational feasibility (e. g., use offloading to allow more computational intensive tasks if latency is not a problem but always favor the decentralized deployment) and (b) the amount and type of information flow between the system components to increase efficiency (e. g., with respect to the (online) task allocation problem). The latter also considers the question of what information needs to be stored (and where).
2. Multiple robots and their coordination constitutes the technical and methodological base, and CPPS require adequate concepts for navigation (path planning and localization) and collision handling. Possible solutions need to consider practical problems like network delays, localization inaccuracies, and the availability of computational power (tailored to embedded systems).

We believe our system design encourages the integration of such aspects into a holistic approach of applying agents in CPPS and, finally, allows the verification and determination of such effects in order to assess its reasonableness.

Finally, it should be noted that further improvements of the score function (see Eqn. 1) are planned and, once the framework evolves further, efficiency and robustness will be evaluated.

Acknowledgments. This work was supported by the German Federal Ministry for Economic Affairs and Energy (BMWi) under the "AUTONOMIK für Industrie 4.0" research program within the SMART FACE project (Grant no. 01MA13007).

References

1. Bader, M., Richtsfeld, A., Suchi, M., Todoran, G., Holl, W., Vincze, M.: Balancing Centralised Control with Vehicle Autonomy in AGV Systems for Industrial Acceptance. In: 11th Int. Conf. on Autonomic and Autonom. Sys.) (2015)
2. Bochmann, L., Gehrke, L., Böckenkamp, A., Weichert, F., et al.: Towards Decentralized Production: A Novel Method to Identify Flexibility Potentials in Production Sequences Based on Flexibility Graphs. Int. J. of Autom. Tech. 9(3) (2015)
3. Digani, V., Sabattini, L., Secchi, C., Fantuzzi, C.: Towards Decentralized Coordination of Multi Robot Systems in Industrial Environments: A Hierarchical Traffic Control Strategy. In: IEEE Int. Conf. on Intelligent Comp. Comm. and Proc. pp. 209–215 (2013)
4. Hart, P., Nilsson, N., Raphael, B.: A Formal Basis for the Heuristic Determination of Minimum Cost Paths. IEEE Trans. on Systems Science and Cybernetics 4(2), 100–107 (July 1968)
5. Hennes, D., Claes, D., Meeussen, W., Tuyls, K.: Multi-robot Collision Avoidance with Localization Uncertainty. In: Proc. of the 11th Int. Conf. on Autonomous Agents and Multiagent Systems. AAMAS '12, vol. 1, pp. 147–154 (2012)
6. Jazdi, N.: Cyber physical systems in the context of Industry 4.0. In: IEEE Int. Conf. on Aut., Quality and Testing, Robotics. pp. 1–4 (May 2014)
7. Kallmann, M.: Shortest Paths with Arbitrary Clearance from Navigation Meshes. In: Proc. of the Eurographics / SIGGRAPH Symp. on Comp. Animation (2010)
8. Kallmann, M.: Dynamic and Robust Local Clearance Triangulations. ACM Trans. Graph. 33(5), 161:1–161:17 (Sep 2014)
9. Otte, M., Correll, N.: The Any-Com Approach to Multi-Robot Coordination. In: Proc. 10th Int. Symp. on Dist. Auton. Rob. Sys. (2010)
10. Purwin, O., D'Andrea, R., Lee, J.W.: Theory and Impl. of Path Planning by Negotiation for Decentralized Agents. Robot. Auton. Syst. 56(5), 422–436 (May 2008)
11. Quigley, M., Conley, K., Gerkey, B., Faust, J., Foote, T., Leibs, J., Wheeler, R., Ng, A.: ROS: An Open-Source Robot Operating System. In: ICRA Workshop on Open Source Software (2009)
12. Thrun, S., Burgard, W., Fox, D.: Probabilistic Robotics (Intelligent Robotics and Autonomous Agents). MIT Press (2005)
13. Weyns, D., Holvoet, T., Schelfthout, K., Wielemans, J.: Decentralized Control of Automatic Guided Vehicles: Applying Multi-Agent Systems in Practice. In: 23rd ACM SIGPLAN Conf. on Object-oriented Programming Sys. Languages and Applications. pp. 663–674. OOPSLA Companion, ACM (2008)

Printed in the United States
By Bookmasters

Printed in the United States
by Bookmasters